In-Situ Rock Stress - Measurement, Interpretation and Application

In-Situ Rock Stress - Measurement, Interpretation and Application

Editor

Abhinav Mishra

In-Situ Rock Stress - Measurement, Interpretation and Application
Edited by **Abhinav Mishra**

Printed in 2017

ISBN: 978-1-68117-404-4

Library of Congress Control Number: 2015941600

© 2016 by ˙
SCITUS Academics LLC,
616, Corporate Way, Suite 2, 4766,
Valley Cottage, NY 10989

www.scitusacademics.com

Contents

Preface

Understanding in-situ rock stress is important in the exploration and engineering involving rock masses for mining, hydropower, tunneling, oil and gas production, and stone quarrying. Traditional methods of determining these stresses have not developed substantially to keep pace with the increasing utilization of rock masses. Contributed by a group of leading experts, this book addresses new developments in numerical modeling and advanced measuring techniques. In-Situ Rock Stress: Measurement, Interpretation and Application reflect the development in this field, covering measuring techniques, interpretation methods, and application of the in-situ stress in engineering practice. Estimate of the in-situ rock stress state can be realized by direct or indirect methods. Although the indirect method has developed rapidly in recent years, the direct field measurement is still by far dominating. Great improvements have been achieved with the 'traditional' field tests by overcoring and hydraulic fracturing, whilst the recently developed methods become matured. In addition, ideas of new methods and new instruments will make the stress estimate easier, less expensive and more reliable.

Editor

Physical Property Relationships of the Rotokawa Andesite, a Significant Geothermal Reservoir Rock in the Taupo Volcanic Zone, New Zealand

Paul A Siratovich[1], Michael J Heap[2], Marlène C Villenueve[1], James W Cole[1], and Thierry Reuschlé[2]

[1]Department of Geological Sciences, University of Canterbury, Private Bag 4800, Christchurch 8140, New Zealand

[2]Laboratoire de Déformation des Roches, Équipe de Géophysique Expérimentale, Institut de Physique de Globe de Strasbourg (UMR 7516 CNRS, Université de Strasbourg/EOST), 5 rue René Descartes, Strasbourg cedex 67084, France

ABSTRACT

Background

Geothermal systems are commonly hosted in highly altered and fractured rock. As a result, the relationships between physical properties such as strength and permeability can be complex. Understanding such properties can assist in the optimal utilization of geothermal reservoirs. To resolve this issue, detailed laboratory studies on core samples from active geothermal reservoirs are required. This study details the results of the physical property investigations on Rotokawa Andesite which hosts a significant geothermal reservoir.

Methods

We have characterized the microstructure (microfracture density), porosity, density, permeability, elastic wave velocities, and strength of core from the high-enthalpy Rotokawa Andesite geothermal reservoir under controlled laboratory conditions. We have built empirical relationships from our observations and also used a classical micromechanical model for brittle failure. Further, we compare our results to a Kozeny-Carman permeability model to better constrain the fluid flow behavior of the rocks.

Results

We show that the strength, porosity, elastic moduli, and permeability are greatly influenced by pre-existing fracture occurrence within the andesite. Increasing porosity (or microfracture density) correlates well to a decreasing uniaxial compressive strength, increasing permeability, and a decreasing compressional wave velocity.

Conclusions

Our results indicate that properties readily measurable by borehole geophysical logging (such as porosity and acoustic velocities) can be used to constrain more complex and pertinent properties such as strength and permeability. The relationships that we have provided can then be applied to further understand processes in the Rotokawa reservoir and other reservoirs worldwide.

BACKGROUND

Fractures on multiple scales are the dominant control on fluid flow in most geothermal systems worldwide. Geothermal environments are prone to variable heat fluxes, dynamic fluid flow regimes, and active tectonics which impact the physical and mechanical properties of the reservoir rocks in which they are hosted. The influence of such a dynamic environment can render the host rocks highly altered, fractured, and microstructurally complex. As a result, the empirical correlation of physical properties to yield valuable relationships may not be entirely straightforward. Studies of these properties, and attempts to quantify how they relate to one another in the subsurface, can greatly assist in the optimization and maintenance of geothermal resources (e.g., Gupta and Sukanta [2006]; DiPippo [2008]; Grant and Bixley [2011]).

Here, we detail the results of a systematic physical and mechanical property study on the Rotokawa Andesite; the major reservoir unit within the high-enthalpy Rotokawa Geothermal Field (Krupp and Seward [1987]; Quinao et al. [2013]), located within the Taupo Volcanic Zone (TVZ), North Island, New Zealand (Figure 1). We first examine the texture, mineralogy, petrology, and microstructure. The key physical properties are then explicitly investigated: porosity, density, elastic wave propagation and dynamic elastic moduli, uniaxial compressive strength, static elastic moduli, and permeability. We empirically correlate the microcrack density of the andesite to the measured physical properties. Further, we present empirical relationships of physical properties and classical micromechanical and geometrical models to predict both uniaxial compressive strength and permeability, respectively. Our data is discussed in relation to the Rotokawa Geothermal Field and their applicability to other geothermal resources worldwide.

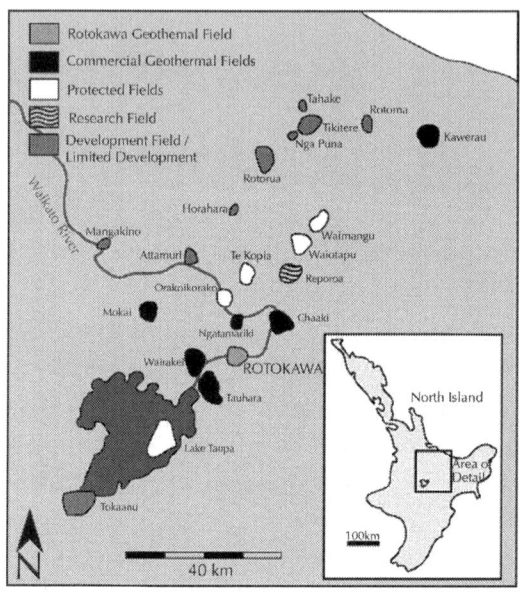

Figure 1: Geothermal fields of the Taupo Volcanic Zone (TVZ), North Island, New Zealand. Core used for this study was sourced from wells drilled in the Rotokawa Geothermal Field. (Adapted from Sewell et al. [2012]).

Previous Studies of Relevance

The study of the core from geothermal systems can yield valuable information to assist their modeling and understanding. For example, Stimac et al. ([2004]) present a study detailing the relationship between permeability and porosity from continuous core from Tiwi geothermal field, Philippines. Their data show that permeability and porosity decrease with depth, with occasional deviations attributed to alteration and compaction. However, the authors are careful to note that their work does not consider the influence of microfractures and their effect on relevant reservoir parameters. Lutz et al. ([2010]) present a case history of the well core from the Desert Peak field (NV, USA) in preparation for the stimulation of an enhanced geothermal system (EGS) by a thorough evaluation of petrological strength and elastic moduli. The results of their study elucidate relationships between clay mineralogy, rock fabric, and permeability increases as a result of mechanical shearing which support proposed hydraulic fracture operations in Well 27-15 at Desert Peak.

The effect of hydrothermal alteration on the physical properties of geothermal core is also a very significant area of research. Hydrothermal alteration can drastically change the elastic wave velocities and permeabilities of rock in both the natural and laboratory environment (Jaya et al.[2010]; Kristinsdóttir et al. [2010]; Pola et al. [2014]). However, coupled studies of physical properties such as porosity, permeability, and strength on geothermal reservoir rocks have not been extensively presented. A detailed study of the impact of a complex microstructure (microfractures and hydrothermal alteration) on the rock physical properties of a geothermal system such as Rotokawa could serve to greatly improve the understanding of reservoir processes at multiple scales.

Geothermal systems are more often than not associated with volcanic systems and are often hosted in rocks sourced from extinct volcanic systems. By proxy, the study of rocks from volcanic edifices can help to boost the understanding of processes within geothermal reservoirs especially with regard to microfractures, which play an essential role in controlling strength, porosity, permeability, elastic wave velocities, and elastic moduli of rocks (Wu et al. [2000]; Guéguen and Schubnel [2003]; Pereira and Arson [2013]; Faoro et al. [2013]; Pola et al. [2014]; Heap et al.[2014]). For example, Vinciguerra et al. ([2005]) studied the influence of thermal stressing on basaltic samples. They show, using elastic wave velocities, that the response of microstructurally variable basalts to thermal stressing can be quite different. While fresh microlitic basalt exhibited severe reductions in P-wave velocity after exposure to 900°C, the P-wave velocity of porphyritic basalt with a pervasive microcrack network did not change.

Similar dependence on the effect of microfractures on strength (Smith et al. [2009]) and permeability (Nara et al. [2011]) has been investigated, with microfractures proving to be deleterious to strength and to enhance permeability. Heap et al. ([2014]) showed, for a suite of pervasively fractured andesites, that an increase in porosity from 8 to 29 vol% decreases strength by a factor of 8 and increases permeability by 4 orders of magnitude. David et al. ([1999]) showed that mechanical and thermal microcracking in granites results in significant changes to permeability and elastic wave velocities. Mechanical microcracking resulted in the development of P-wave velocity anisotropy, while thermally microcracked samples showed little P-wave anisotropy. Additionally, permeability was much more varied in mechanically microcracked rocks than those induced thermally, suggesting that thermal microcracks develop isotropically. Chaki et al. ([2008]) investigated the role of thermal microcracking in granites and showed that elastic wave propagation is attenuated by microcracks and the orientation of these thermal microcracks (with regard to the original microstructure) plays a critical role in the propagation and attenuation of the waves. Faoro et al. ([2013]) provide a model for how microcrack density within an isotropically microcracked sample can be modeled as a function of aspect ratio and microcrack connectivity. Elastic moduli and elastic wave velocities are strongly influenced by the morphology, distribution, and shape of pore space in rocks and are substantially attenuated by the presence of microcracks (Stanchits et al. [2006] and references therein).

The relationship between porosity and strength has been observed by many authors, with general agreement that as the porosity of a sample (both rock and other engineering materials) increases, the strength decreases (e.g., Al-Harthi et al. [1999]; Li and Aubertin [2003]; Kahraman et al.[2005]; Chang et al. [2006]; Diamantis et al. [2009]; Ju et al. [2013]; Baud et al. [2014]; Heap et al. [2014]). The geometry of the pores also has a significant role in the strength of the materials both intrinsically and with respect to the direction of stress (Luping [1986]). The microstructure of rocks can be changed by increasing the crack damage (by mechanical and/ or thermal stresses) as well as hydrothermal alteration (Heap et al. [2009]; Nara et al. [2011]; Pola et al. [2014]); these changes can be observed through the evaluation of destructive and nondestructive physical property measurements (Pola et al. [2012] and references therein; Sousa et al. [2005]). Further, Pola et al. ([2014]) also show that hydrothermal alteration of volcanic rocks can either strengthen or weaken rocks by decreasing or increasing their porosity, respectively.

Geological Significance of the Rotokawa Andesite

The TVZ is a rifted arc associated with the Hikurangi subduction system in which the Pacific plate descends beneath the Australasian plate (Cole [1990]; Wilson et al. [1995]), and hosts active volcanism and multiple associated hydrothermal systems (Bibby et al. [1995]; Rowland and Sibson [2004]; Rowland et al. [2010]). The Rotokawa field is one of these active hydrothermal systems and has been the subject of exploration for mineral resources (sulfur and gold deposits) and, for many years, was the subject of detailed investigation into its use as a commercial geothermal resource (Collar and Browne [1985]; Krupp and Seward [1987]; Hedenquist et al. [1988]). More recently, electricity generation has been realized at Rotokawa following the installations of the Rotokawa I (1997) and Nga Awa Purua (2010) generation stations (Legmann and Sullivan [2003]; Bloomberg et al. [2012]). The more recent of these installations, the Nga Awa Purua power station, hosts the single largest geothermal turbine installation in the world and has a generation capacity >140 MWe which is approximately 3 % of New Zealand's electricity consumption (Horie and Muto[2010]).

The main production zone for the installations at Rotokawa is from that of the Rotokawa Andesite, a series of lavas, pseudo-breccias, and breccias. The movement of fluid through the andesite is predominantly along fracture networks (Rae [2007]; Massiot et al. [2012]). The andesite overlies basement of Miocene greywacke and is capped by a sequence of volcaniclastic and sedimentary units: Reporoa Group, Wairakei Ignimbrite, Waiora Formation, and Huka Falls Formation (Krupp and Seward [1987]; Rae [2007]). The andesite is gray to green and occasionally purple in color, depending on alteration within the reservoir; alteration is less intense in the lavas and more intense in the breccia and pseudo-breccia (Ramirez and Hitchcock [2010]). Production of reservoir fluids is sourced from the Rotokawa Andesite by 12 wells in the central part of the field (Figure 2), and re-injection of spent fluids is done through 5 wells along the southeastern margin of the field (Powell [2011]).

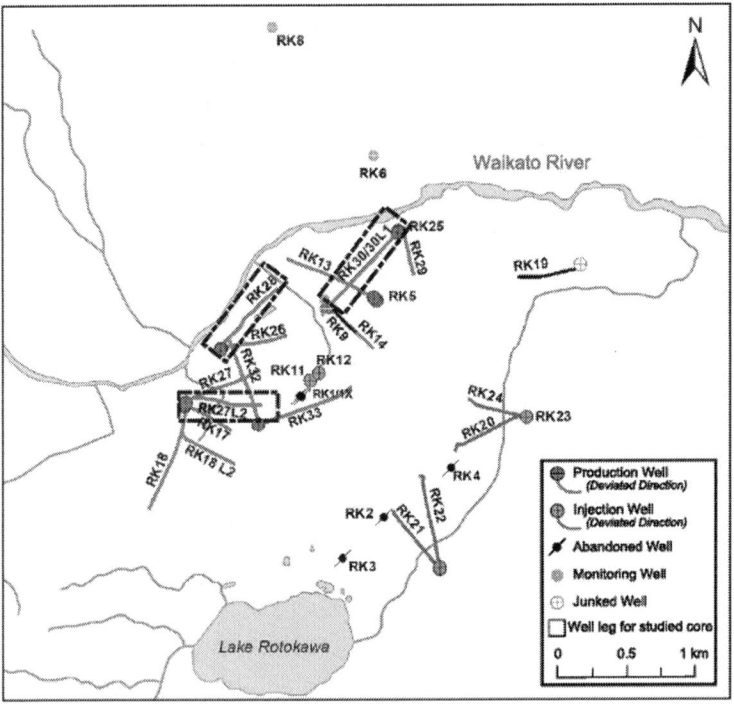

Figure 2: Rotokawa geothermal field and production and injection wells used within the field. Wells RK27L2, RK28, and RK30 were the source of the core used in this study and are outlined with dashed boxes in the figure.

METHODS

Study Source Material

The samples used in this study were sourced from Rotokawa production wells RK27L2, RK28, and RK30 (Figure 2). The measured depths (total borehole length measured from drilling rig floor), orientations, and corrected true vertical depths (TVD) are listed in Table 1. The original cores were approximately 6-m long and 100-mm diameter, and were initially described by the GNS Science Wairakei Research Centre, New Zealand, in a series of internal industry reports detailing the respective production wells from surface to total depth (TD). These reports describe the stratigraphic sequences of the wells and rock types, hydrothermal alteration, and locations of the wells (Rae et al. [2009]; Rae et al. [2010]; Ramirez and Hitchcock [2010]).

Table 1: Detail of core retrieval points from within the Rotokawa Andesite reservoir

Well name	Measured depth of core points (m)	True vertical depth (meters below reference level)	Inclination from vertical (degrees)	Azimuth from north (degrees)
RK28 ST1	2,310 to 2,316	−2,215 to 2,221	21.94	50.91
RK27 L2	2,120 to 2,126	−2,001 to 2,007	27.31	88.11
RK30 L1	2,320 to 2,326	−2,175 to 2,182	20.11	218.47

Measured depths are given as measured from the elevation of the drilling rig floor, true vertical depths are corrected to subsurface elevations, inclination is the deviation from vertical, and azimuth is the orientation of the borehole trajectory.

Siratovich et al.

Siratovich et al. Geothermal Energy 2014 2:10   doi:10.1186/s40517-014-0010-4

At the University of Canterbury (UC), the cores were catalogued and cut into workable cylinders approximately 100 mm in length. These smaller sections were over-cored to obtain smaller cylindrical samples 40 mm in diameter and ranging from 80 to 100 mm in length. All samples were machined so that their end faces were flat and parallel in accordance with ISRM standards (Ulusay and Hudson [2007]).

Microstructural Characterization

The strength, porosity, permeability, and acoustic velocities can be significantly influenced by the presence of microfractures in a sample. Therefore, we deemed it necessary to develop a fundamental understanding of the microfracture densities in the samples. In order to characterize these features, 10 polished thin sections were prepared from offcuts of the cylinders used for the property characterization described below. The thin sections were prepared perpendicular to the core axis (X-Y plane) and, using reflected light thin section photomicrography (at ×40 magnification), were examined for microfracture densities using the methods suggested by Underwood ([1970]) and further described by Richter and Simmons ([1977]), Wu et al. ([2000]), and Heap et al. ([2014]). In each thin section, an 11×11 mm^2 area was selected, which was subdivided into sections of 1×1 mm^2. The number of cracks that intersected a grid array of parallel and perpendicular lines that were spaced at 0.1 mm was counted. This allowed the calculation of the crack surface area per unit volume according to Equation 1 (Underwood [1970]):

$$Sv = \pi/2 P_\mathrm{I} + (2 - \pi/2) P_\mathrm{II}$$

(1)

where Sv is the crack surface area per unit volume (in mm^2/mm^3), P_I is the number of perpendicular lines crossed by crack intersections, and P_II is the number of parallel lines crossed by crack intersections. We also characterized the anisotropy of microfracture distribution using Equation 2 (Underwood [1970]):

$$\Omega_{23} = P_\parallel - P_\parallel / P_\parallel + (4/\pi - 1)P_\parallel \qquad (2)$$

Density and Porosity Measurements

Once the samples were cut and ground flat and parallel, they were washed with water to remove any debris from sample preparation. They were then immersed in distilled water under vacuum of about 100 kPa for 24 h. Samples were taken out of the water and were weighed after their surface water had been removed. The samples were then placed into a laboratory oven at 105°C and dried until a constant mass was observed. Subsequently, they were removed from the oven and held in a dessicator until further characterization was implemented. Sample lengths and diameters were measured to within 0.01 mm. The connected porosity and dry bulk density of the samples were calculated following the methods recommended by Ulusay and Hudson ([2007]).

Characterization of Elastic Wave Velocities and Dynamic Elastic Moduli

The compressional wave (Vp) and shear wave (Vs) velocities and dynamic elastic moduli were measured using a GCTS (Geotechnical Consulting and Testing Systems, Tempe, AZ, USA) Computer Aided Ultrasonic Velocity Testing System (CATS ULT-100) apparatus with axial P- and S-wave piezoelectric transducers (Figure 3). The resonance frequency of the transducers was 900 kHz, pulse acquisition rate was 20 MHz, and 108 waveforms were captured for each sample. The velocities were collected under a constant uniaxial stress of 10 MPa via a Tecnotest servo-controlled 3,000 kN loading frame (Technotest, Modena, Italy) (Figure 3). The stress of 10 MPa was used to ensure a consistent waveform across the specimens and that applied stress was consistent for all measurement cycles. This was determined to be below microcrack closure and opening stress by analyzing the change in axial strain as the sample was loaded to 10 MPa (Eberhardt et al. [1998]). There was no change in axial strain and absence of acoustic emissions (AEs) during the initial loading (Brace et al. [1966]; Martin and Chandler [1994]; Lion et al. [2005]; Nicksiar and Martin [2012]); this ensured a good quality interpretation of the first arrival time of elastic wave pulses. Using these data, we determined the dynamic Poisson's ratio and Young's modulus using Equations 3 and 4 (Guéguen and Palciauskas [1994]), respectively:

$$V_d = \left(Vp^2 - 2Vs^2\right)/2\left(Vp^2 - Vs^2\right) \tag{3}$$

$$E_d = \left(\rho Vs^2 \left(3\,Vp^2 - 4Vs^2\right)\right)/\left(Vp^2 - Vs^2\right) \tag{4}$$

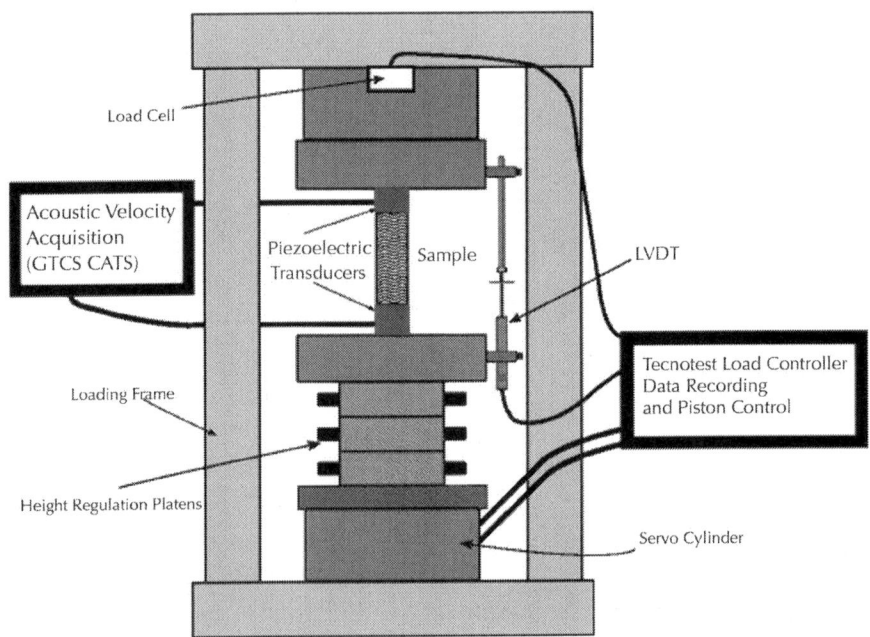

Figure 3: Loading frame set-up for acquisition of elastic velocities at the University of Canterbury. The frame is a Tecnotest 3,000 kN loading frame and a stress of 10 MPa was applied to each sample cycle to ensure a consistent waveform and quality picking of the first arrival time of the acoustic pulses (note that the figure is not to scale).

Where Vp is compressional wave velocity in meters per second, Vs is shear wave velocity in meters per second, E_d is the dynamic Young's modulus in pascal, V_d is the dynamic Poisson's ratio, and ρ is density in kilograms per cubic meter. Our physical property characterizations are summarized in Table 2. In addition to the determination of the elastic moduli from our elastic wave velocity measurements, we also utilized the method of Martínez-Martínez et al. ([2011]) to characterize the spatial attenuation of the compressional waveform anisotropy (Equation 5):

$$\alpha_s = 20 \log \left(A_e / A_{max} \right) / L$$

(5)

where α_s is spatial attenuation in decibels per centimeter, A_e is the maximum amplitude emitted by the piezoelectric crystal, A_{max} is the recorded maximum amplitude of the pulse after passing through the sample, and L is the length of the sample in meters.

Table 2: Results of quantitative microstructural characterization

Sample name	Crack density for intercepts parallel to orientation axis $P \parallel$ (mm−1)	Crack density for intercepts perpendicular to orientation axis $P \perp$ (mm−1)	Crack area per unit volume Sv (mm2/ mm3)	Anisotropy factorΩ2,3	Connected porosity (vol%)
27_21_0B	13.73	13.53	13.06	0.01	14.91
28_10_5A	4.10	4.41	8.33	0.06	7.47
27_20_4_B	1.83	2.06	3.77	0.08	4.37
27_3_3B	4.77	4.77	9.55	0.01	9.81
30_22_4A	2.77	2.62	5.48	0.01	6.49
28_10_9B	3.84	3.90	7.71	0.03	7.42
28_12_1	4.83	4.65	9.59	0.02	7.89
30_21_1B	2.93	2.91	5.85	0.01	7.47
27_21_3A	5.80	5.13	11.31	0.09	16.3
28_10_6C	2.97	2.87	6.38	0.05	5.97

As discussed in the section on microstructural characterization, crack densities were calculated on thin section samples to ascertain crack areas per unit volume using the optical microscope method (Underwood [1970]).

Siratovich *et al.*

Siratovich *et al. Geothermal Energy* 2014 2:10   doi:10.1186/s40517-014-0010-4

Uniaxial Compressive Strength testing and Static Elastic Moduli

Uniaxial compressive strength (UCS) was determined using a Technotest 3,000 kN, servo-controlled loading frame (Figure 4). Four strain gauges (20-mm strain gauges with a gauge factor of 2.12 supplied by Tokyo Sokki Kenkyujo Co. Ltd. (TML) Shinagawa-ku, Tokyo, Japan) were glued onto each sample. Two vertical gauges measured axial strain and two laterally oriented gauges measured radial strain; care was taken to ensure that the strain gauges were perpendicular to their respective axes of deformation. The specimens were deformed at a constant strain rate of 1.0×10^{-5} s^{-1} (controlled by linear variable differential transformer, LVDT) at ambient laboratory temperature and humidity conditions. During experimentation, AE output was monitored using Physical Acoustics Corporation MISTRAS' AE node acquisition system (Princeton Jct, NJ, USA). Two physical acoustics WS AE transducers (100 to 900 kHz operating frequency) were attached to the samples at the top and base, and hit counts, waveforms, energy, and amplitude of the received signals were recorded during sample deformation. AE monitoring was used during deformation as a proxy for microcracking as AEs are generated by the release of energy from a material during the propagation and nucleation of microcracks (Eberhardt et al. [1998]; [Diederichs et al. 2004]). We utilized arbitrary AE energy units (the area under the received waveform signal) for comparison of AE activity across the datasets. Once stress-strain curves were obtained and AE data are processed, we calculated the static elastic moduli for each specimen utilizing Equations 6 and 7 with the tangent deformation modulus at 50 % of the maximum peak stress (Ulusay and Hudson [2007]). In addition, we selected portions of the stress-strain sequence to identify crack closure, crack initiation, unstable crack propagation, and, ultimately, crack coalescence and sample failure (Martin [1993]; Eberhardt et al. [1998]; Takarli et al. [2008]; Heap and Faulkner[2008]):

$$E_s = (\Delta\sigma a / \Delta\varepsilon a) \tag{6}$$

$$v_s = -(\Delta\varepsilon r / \Delta\varepsilon a) \tag{7}$$

where E_s is the static Young's modulus (Pa), v_s is the static Poisson's Ratio, σa is the differential axial stress (Pa), εa is the axial strain, and εr is the radial strain.

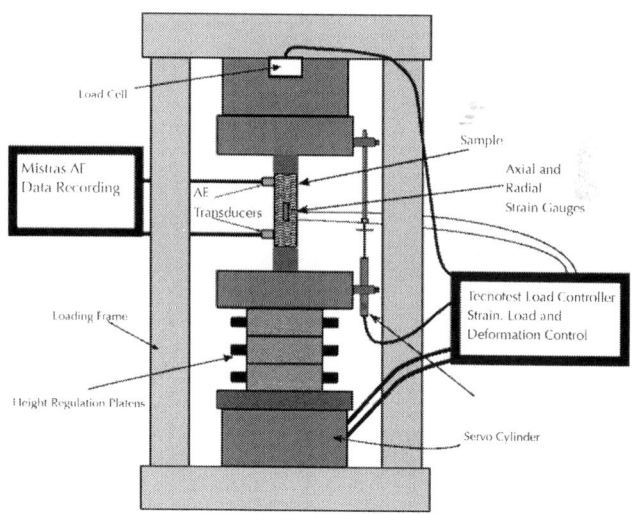

Figure 4: Loading frame set-up for determining uniaxial compressive strength (UCS) at University of Canterbury. Testing was carried out using TML strain gauges and MISTRAS acoustic emission monitoring equipment. Loading during axial differential stressing was achieved through a constant strain rate of 1.0×10^{-5} s^{-1} (not to scale).

Permeability Measurements

Gas (argon) permeability measurements were made at the Laboratoire de Déformation des Roches, Université de Strasbourg (France). The 40-mm diameter specimens were over-cored to a diameter of 20 mm and cut and ground flat and parallel to a nominal length of 40 mm. The new samples were then re-evaluated by the triple-weight method to obtain porosity via the Archimedes' method (Ulusay and Hudson [2007]) and oven-dried under vacuum at 40°C until no change in sample mass was observed. The samples were then jacketed with viton sleeves, placed between two steel end-caps and lowered into the pressure vessel (Figure 5). A confining pressure of 2 MPa was applied to the sample (provided by distilled water), and permeability measured using the transient method (or pulse-decay method). For the permeability measurements, an initial differential pore pressure was applied to the sample, the upstream inlet was then closed, and the pore pressure decay monitored over time. The downstream fluid pressure (P_{down}) was the ambient atmospheric pressure, and the maximum upstream fluid pressure (P_{up}) was set so the pressure differential was 0.5 MPa. Permeability was then calculated using Equation 8 (after Brace et al.[1968]):

$$k_{gas} = (2\eta L/A)\left(V_{up}/\left((P_{up})^2 - (P_{down})^2\right)\right)(\Delta P_{up}/\Delta t) \tag{8}$$

where k_{gas} is the gas permeability, η is the viscosity of the pore fluid, A is the cross-sectional area of the sample, V_{up} is the volume of the upstream pore pressure circuit (approximately 7 cm³), P_{up} is the upstream pore pressure, P_{down} is the downstream pore pressure, and t is the time. By plotting ΔP_{up} as a function of time, the local slope of the curve is computed to determine the temporal variation of the permeability k_{gas}. To check whether our data should be corrected for Klinkenberg's 'slip flow' (Klinkenberg [1941]), we plotted the measured gas permeability as a function of the inverse of the mean pore fluid pressure, P_{mean}. For the transient method, since P_{down} is constant, the decay of P_{up} through time corresponds to the decay of the mean pore pressure P_{mean}. We found that, in all cases, the Klinkenberg correction should be applied:

$$k_{\text{true}} = k_{\text{gas}}\left(1 + b / P_{\text{mean}}\right)$$

(9)

where k_{true} is the true gas permeability, b is Klinkenberg slip factor, and P_{mean} is the mean pore fluid pressure.

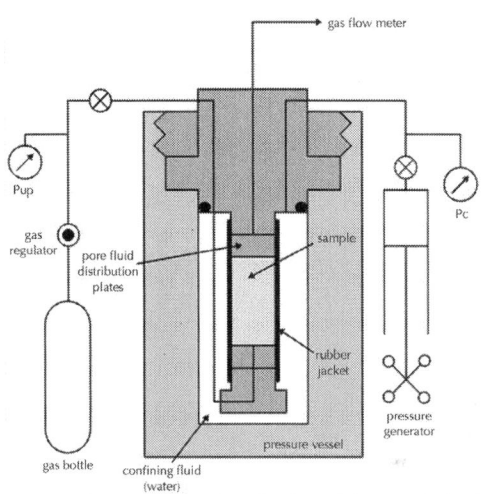

Figure 5: Gas permeameter used to measure permeability at University of Strasbourg (not to scale). Confining pressure of 2 MPa was applied using distilled water as the pressurizing media, and pore fluid was dry argon gas supplied at 1.5 MPa for a net effective pressure of 0.5 MPa.

RESULTS

In the following section, we present our data and observations on petrology, microstructure (quantitative microfracture analysis), macrostructure (bulk density, porosity, acoustic wave velocities, and dynamic moduli), strength relations (by UCS testing), and finally the ability of the rock to transmit fluid (permeability) of the Rotokawa Andesite.

Petrology

The Rotokawa Andesite shows moderate to intense hydrothermal alteration with the groundmass and phenocrysts showing replacement of original mineralogy. Fractures and occasional veins of quartz, calcite, anhydrite, and epidote occur, and amygdales within the sample are often filled with chlorite, calcite, hematite, pyrite, and chalcedony, often with quartz rims (Figure 6). Alteration is pervasive with the original mineral assemblages typically replaced by secondary hydrothermal alteration species, with some specimens showing very little original mineralogical texture. Plagioclase feldspars have been altered to albite, adularia, occasional calcite, and rare pyrite, and ferromagnesian minerals have been replaced by chlorite, quartz, calcite, and occasional epidote. Microfractured phenocrysts (Figures 6 and 7) are abundant and many relict phenocrysts retain original texture are but replaced by secondary mineralization. The average phenocryst size is 0.5 to 1 mm with occasional plagioclase near 1.5 to 2 mm; amygdales also range from 1 to 1.5 mm in size. The alteration chemistry of the samples indicates that this portion of the reservoir is dominated by chlorite/epidote alteration. The degree of alteration is relatively consistent across the core we have sampled with most primary mineralogies replaced by secondary alteration products. Microfracture mineralization indicates that these networks may have been conductive pathways for fluid migration (i.e., the presence of chlorite clays, adularization of plagioclase, calcite, and quartz rimming of fractured matrix); we typically observe chlorite, calcite, and quartz as alteration mineralogies with occasional epidote centers within the fractures. Backscatter scanning electron microscopy (SEM) was utilized to further reiterate the complex interaction of fractures and vesicles in the specimens (Figure 7). At several different magnifications, we see an abundance of microfractures in the samples as well as a clear depiction of the complex alteration mineralogy displayed by the andesite.

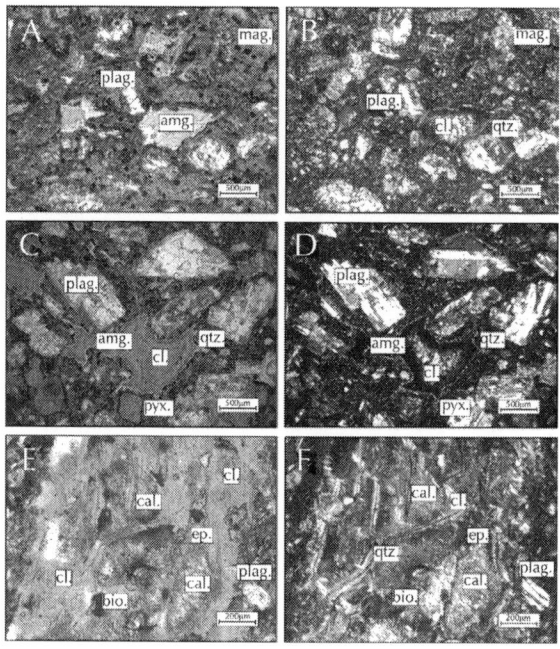

Figure 6: Thin section photomicrographs of the Rotokawa Andesite. (A) Plane-polarized light of RK28 2310.6C andesitic pseudo-breccia with plagioclase laths (plag.), groundmass is altered plagioclase, abundant magnetite (mag.), and amygdale (amg.). (B) Cross-polarized light of RK28 2310.6C clearly shows alteration fabrics of the brecciated andesite with plagioclase (plag.) and amygdales filled with chlorite (cl.) and rimmed by quartz (qtz.). (C) Plane-polarized light view of RK27_L2 2121.4A showing andesitic breccia with plagioclase with slight adularia alteration (plag.) and amygdale (amg.) filled with chlorite (cl.), quartz (qtz.), and highly altered pyroxene (pyx.) in lower portion of image. (D) Cross-polarized light view of RK27_L2 2121.4A shows chlorite infill of a large amygdale (amg.) in the center of the photomicrograph and quartz rim (qtz.), plagio-clase (plag.), and highly altered pyroxene (pyx.). (E) Plane-polarized light view of RK30 2322.4A shows highly altered and microfractured plagio-clase phenocryst with intense alteration and replacement by chlorite (cl.), epidote (ep.), calcite (cal), possible biotite (bio.), and small plagioclase showing evidence of adularia alteration (plag.). (F) Cross-polarized light of RK30 2322.4A illustrates microfracture network and veining with al-teration products of quartz (qtz.), epidote (ep.), biotite (bio.), calcite (cal.), and plagioclase (plag.).

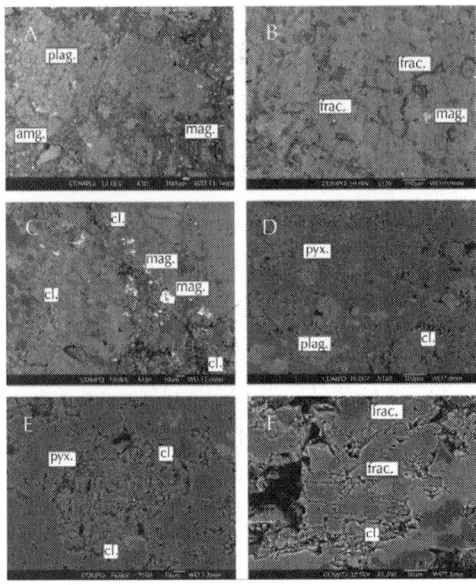

Figure 7: Backscattered scanning electron microscope photomicrographs of Rotokawa Andesite. Samples A to C are from RK28 at 2,310.6 m depth and D to F are from RK27 at 2,121.1 m depth.(A) Andesite with abundant ferromagnesian minerals (mag.), altered amygdale (amg.), and highly scattered magnetites (mag. bright hues).(B) Detail of a fractured plagio-clase phenocryst showing the microfractured texture (frac.) and occasion-al magnetite (mag.). (C)Detail of a fracture infill showing chloritization (cl.) and abundant ferromagnesian minerals (mag.). (D) Groundmass of RK27 sample; pervasive fracturing is not apparent at this magnification but the porous network is quite apparent with pyroxene (pyx.), chlorite (cl.), and plagioclase (plag.). (E) Replacement mineralogy of likely pyrox-ene phenocryst (pyx.) showing abundant chloritization (cl.) and dissolu-tion textures. (F) Detail of relict pyroxene and abundant chlorite (cl.) with abundant microfractures (cl.) apparent in the sample mass.

Quantitative Two-dimensional Microstructural Analysis

We evaluated the microfracture density of 10 specimens as a function of crack surface area per unit volume (Table 2). These samples were selected to represent the range of connected porosities observed within the sample set. We found that the crack area per unit volume in our samples ranges from 3.77 to 13.06 mm²/mm³ and appears to be independent of the alteration and mineralogy of the specimens. The calculated anisotropy factor ($\Omega_{2,3}$), indicates that the microcracks are isotropic (Table 2).

Porosity and Bulk Density

Bulk density decreases as connected porosity increases, as expected for samples of similar composition (Figure 8A). Bulk dry densities of the samples range from 2.29 to 2.65 g/cm³, with a mean value of 2.49 g/cm³. The connected porosities range from 4.37 to 16.3 vol%, with a mean value of 8.44 vol%. While there is some variation in the distribution of pores/vesicles in the samples, we observe that the microcrack density exerts an important control on the porosity and density, as illustrated by the correlation between crack area per unit volume and connected porosity presented as Figure 8B.

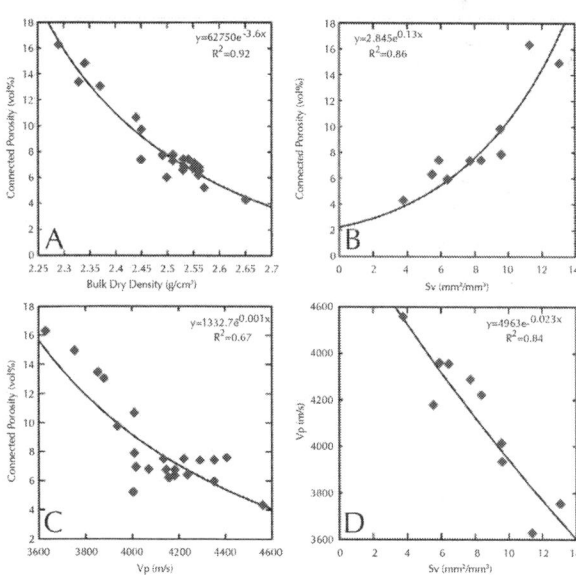

Figure 8: Relationships of porosity, density, crack area, and compressional wave velocity for Rotokawa Andesite. (A) Connected porosity versus

dry bulk density for Rotokawa Andesite calculated using the dual weight method of (Ulusay and Hudson [2007]). (B) Crack area per unit volume (Sv) plotted versus connected porosity for samples representing the range of values measured within the measured dataset.(C) Connected porosity (vol%) using the dual weight method versus axial compressional wave velocity (Vp) as measured under a stress of 10 MPa which was determined to be below the crack closure stress (see text for further detail). (D) Crack area per unit volume (Sv) versus compressional wave velocity (Vp).

Ultrasonic Wave Velocities, Dynamic Elastic Moduli, and Spatial Attenuation

Measurements made on dry samples under ambient (pressure and temperature) conditions yielded axial P-wave velocities from 3,627 to 4,556 m/s with a mean value of 4,106 m/s, and axial S-wave velocities between 2,160 to 2,752 m/s with a mean value of 2,510 m/s (Table 3). Porosity and P-wave velocities show moderate correlation: samples with higher porosities have slower elastic wave velocities, as seen in Figure 8C. The crack area per unit volume also correlates well with P-wave velocity (Figure 8D). The axial spatial attenuation for the andesites ranges from 8.39 to 28.74 dB/cm (Figure 9). Figure 9 shows that there is no clear trend between spatial attenuation and P-wave velocity. Dynamic Poisson's ratio and dynamic Young's modulus were in the range of 0.13 to 0.23 and 24.6 to 45.9 GPa, respectively (Table 3).

Table 3: Physical property measurements of 22 samples used in destructive testing of Rotokawa Andesite

Sample source_well sample name	Bulk dry density (g/cm3)	Connected porosity (vol%)	Vp (m/s)	Vs (m/s)	Spatial attenuation (dB/cm)	UCS (MPa)	Static Young's modulus (GPa)	Dynamic Young's modulus (GPa)	Static Poisson's ratio	Dynamic Poisson's ratio
RK_27_L2_21.5B	2.44	10.72	4,005	2,443	14.63	85.99	19.9	35.1	0.24	0.2
RK_27_L2_21.8A	2.33	13.49	3,850	2,363	16.47	79.91	25.2	31.2	0.26	0.2
RK_27_L2_23.2A	2.56	6.61	4,182	2,490	22.48	105.26	31.2	38.9	0.19	0.23
RK_27_L2_20.4B	2.65	4.37	4,556	2,752	22.53	211.05	37.7	45.9	0.25	0.21
RK_27_L2_21.1C	2.37	13.1	3,877	2,405	23.6	69.53	21.5	32.5	0.18	0.19
RK_27_L2_3.3B	2.45	9.81	3,937	2,331	23.51	95.78	32.4	29.9	0.13	0.18
RK_27_L2_21.0B	2.34	14.91	3,752	2,337	8.4	60.13	28.1	30.6	0.12	0.17
RK_27_L2_21.3A	2.29	16.3	3,627	2,160	15.13	70.57	30.4	24.6	0.16	0.17
RK_28_10.6C	2.5	5.97	4,350	2,652	18.27	146.2	43.7	42.4	0.27	0.2
RK_28_10.8C	2.53	6.72	4,147	2,537	11.67	109.91	27.2	39.1	0.34	0.2
RK_28_10.9B	2.51	7.42	4,285	2,615	14.67	137.31	32.4	41.5	0.2	0.2
RK_28_13.2A	2.55	6.97	4,013	2,531	18.53	146.21	38.3	37.2	0.24	0.19

RK_28_10.5A	2.45	7.47	4,220	2,578	14.83	130.71	27.4	38.8	0.25	0.21
RK_28_11.5A	2.51	7.62	4,403	2,555	10.06	152.76	35.6	37.4	0.27	0.13
RK_28_12.1	2.49	7.89	4,010	2,460	12.23	115.01	29.3	36.2	0.22	0.14
RK_30_20.4A	2.57	5.3	4,002	2,495	10.39	140.97	33.6	36.9	0.09	0.15
RK_30_21.0A	2.55	6.84	4,070	2,508	11.17	126.53	26.4	39.8	0.14	0.15
RK_30_21.1B	2.54	7.47	4,352	2,659	28.75	157.93	25.8	43.2	0.17	0.2
RK_30_21.7B	2.56	6.28	4,154	2,588	19.91	162.71	28.3	41.1	0.22	0.18
RK_30_22.3B	2.53	7.51	4,133	2,550	21.72	137.97	31.5	39.2	0.23	0.19
RK_30_22.4A	2.56	6.49	4,181	2,582	20.27	148.44	34.4	40.7	0.18	0.19
RK_30_22.5B	2.56	6.41	4,236	2,628	23.43	150.71	33.6	42.1	0.09	0.16
Mean	2.49	8.44	4,106	2,510	17.39	124.62	30.6	37.5	0.2	0.18
Standard deviation	0.09	3.23	221	133	5.512	37.01	5.5	5.1	0.06	0.03

Siratovich et al.

Siratovich et al. *Geothermal Energy* 2014 2:10   doi:10.1186/s40517-014-0010-4

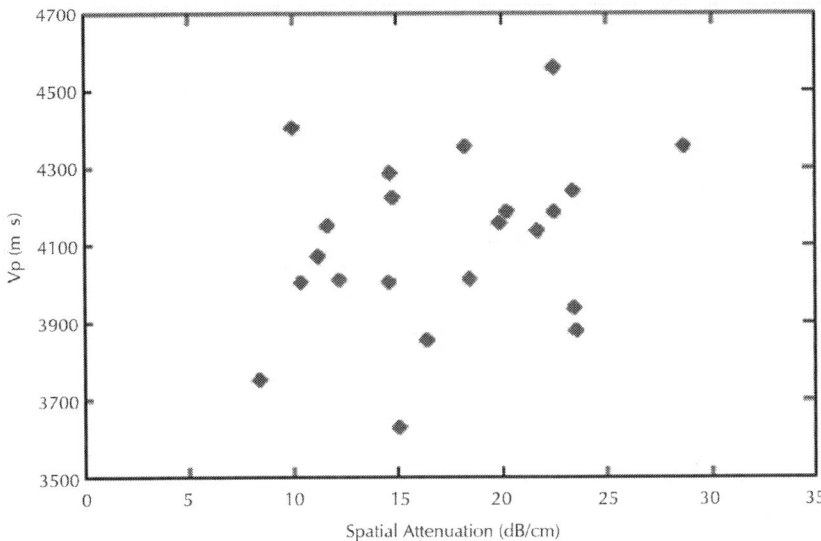

Figure 9: Spatial attenuation (αs) of axial compressional P-wave velocity of Rotokawa Andesite. Attenuation was calculated by the method suggested by Martínez-Martínez et al. ([2011]) utilizing transmission of the ultrasonic wave and maximum attenuation of the waveform plotted versus axial compressional wave velocity (Vp) obtained under an axial stress of 10 MPa.

Uniaxial Compressive Strength and Static Elastic Moduli

In order to characterize the mechanical behavior of the Rotokawa Andesite, the 22 samples of Table 3 were loaded in compression to failure. The dataset shows a large range of UCS (as observed for the other physical properties), from 60 to 211 MPa. The stress-strain behavior of the andesites is very similar across the range of strengths, as shown in Figure 10, which reports curves that best represent the dataset and behavior of the Rotokawa Andesite under uniaxial compression. All specimens in the dataset show brittle behavior as evidenced by the stress-strain relationships and bolstered by analysis of the AE activity. An increase of AE between dilatancy (σ_{cd}) and failure is a benchmark of brittle failure (e.g., Brace and Bombolakis [1963]; Rutter [1986]; Ashby and Sammis [1990]; Heap and Faulkner [2008]), as seen in Figure 10. Weaker specimens showed lower overall AE energy output than higher strength specimens. Static Young's moduli range between 19.9 and 43.7 GPa and static Poisson's ratio between 0.09 and 0.34. Our data shows that as porosity and crack surface area increases, the UCS of the rock decreases (Figure 11A,B). Further, the UCS increases as axial P-wave velocity increases (Figure 11C).

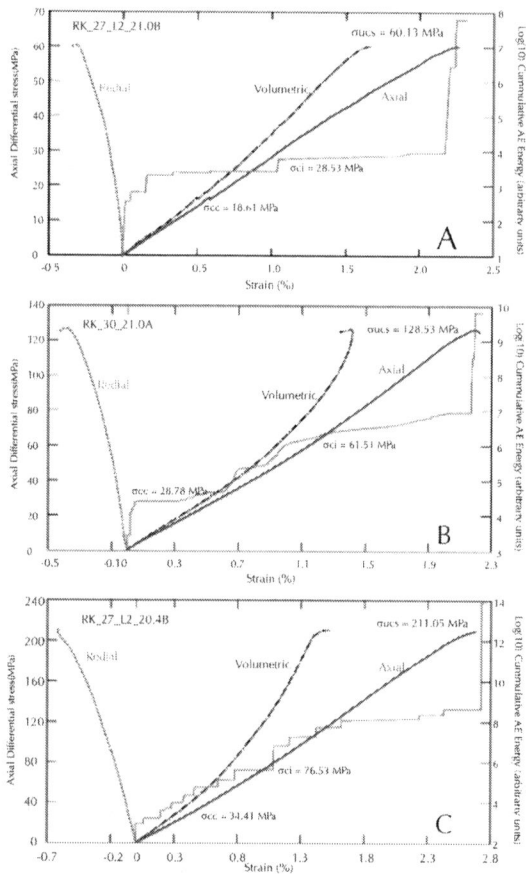

Figure 10: Stress–strain behavior of the Rotokawa Andesite. Samples were subject to constant strain rate loading (1 × 10⁻⁵/s) and monitored for asso-

ciated arbitrary acoustic emission energy output. All samples in this study display brittle failure. (A) Samples with low UCS generally develop a single fracture plane. (B) Samples near mean UCS develop several fracture planes. (C) Samples with very high UCS showed explosive, catastrophic failure into several large and small pieces with no distinct failure plane.

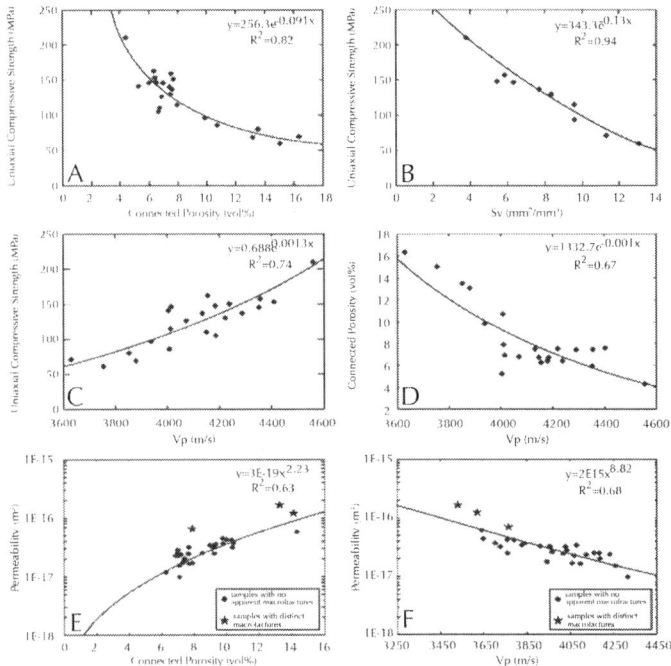

Figure 11: Key empirical relationships of the Rotokawa Andesite. (A) Connected porosity (vol%) plotted versus uniaxial compressive strength (MPa) for Rotokawa Andesite at ambient pressures and temperatures. (B) Crack area per unit volume (Sv) as measured from reflected light thin sections (method recommended by Underwood [1970]) plotted versus uniaxial compressive strength (UCS) for the Rotokawa Andesite. (C) Axial P-wave velocity (Vp) as measured at ambient temperatures under an axial load of 10 MPa versus uniaxial compressive strength values for the Rotokawa Andesite at ambient temperature. (D) Axial compressional wave velocity (Vp) plotted versus connected porosity (vol%) as measured by the dual weight method (Ulusay and Hudson [2007]). (E) Semi-log plot of connected porosity measured by the triple-weight method (Ulusay and Hud-

son [2007]) versus argon gas permeability with effective pressure of 0.5 MPa. (F) Semi-log plot of axial compressional wave velocity (Vp) plotted versus argon gas permeability.

Permeability

Our argon permeability measurements show that for the tested samples, permeability ranges from 9.82×10^{-18} m^2 to 1.66×10^{-16} m^2 (Table 4). The results show a trend of increasing permeability with increasing porosity (Figure 11E). We observe that three of the samples contain macrofractures (black stars on Figure 11E, F) and have higher permeabilities that slightly deviate from the trend of the dataset. We also note that as permeability increases, the axial compressional wave velocity decreases (Figure 11F); further, those samples with distinct macrofractures show lower compressional velocities and higher permeabilities when compared with samples of similar porosity.

Table 4: Results of density, porosity, argon permeability, and acoustic velocity measurements for Rotokawa Andesite

Sample source: well number, depth, name	Bulk dry density (g/cm3)	Connected porosity (vol%)	Argon permeability (m2)	Axial P-wave velocity(m/s)
RK_27_L2_2120.4A_1	2.55	7.24	1.78E−17	3,943
RK_27_L2_2120.4A_2	2.55	7.06	9.82E−18	4,318
RK_27_L2_2121.1A	2.34	13.97	1.25E−16	3,621
RK_27_L2_2123.3A_1	2.48	9.22	3.32E−17	3,911
RK_27_L2_2123.3A_2	2.46	10.25	4.13E−17	3,793
RK_27_L2_2123.7A_1	2.52	9.73	3.69E−17	3,840
RK_27_L2_2123.7A_2	2.50	10.35	3.72E−17	3,704
RK_27_L2_2124.1B_1	2.54	7.63	3.24E−17	3,954
RK_27_L2_2124.1B_2	2.56	6.93	2.81E−17	4,032
RK_27_L2_2121.0A_1	2.42	13.14	1.66E−16	3,532
RK_27_L2_2121.0A_2	2.33	14.23	6.09E−17	3,642
RK_28_2310.3A_1	2.46	9.94	4.26E−17	3,760

RK_28_2310.3A_2	2.48	9.70	4.44E−17	3,650
RK_28_2310.8A_1	2.50	8.95	3.45E−17	3,830
RK_28_2310.8A_2	2.50	9.22	3.43E−17	4,076
RK_28_2310.9C_1	2.46	10.30	3.21E−17	3,725
RK_28_2310.9C_2	2.53	7.64	1.65E−17	4,096
RK_28_2311.1B_1	2.55	7.04	1.53E−17	4,259
RK_28_2311.1B_2	2.56	7.35	2.05E−17	4,190
RK_28_2311.3B_1	2.50	9.10	3.21E−17	4,026
RK_28_2311.3B_2	2.49	9.17	2.50E−17	4,183
RK_30_2320.6A_1	2.53	7.90	7.10E−17	3,765
RK_30_2320.6A_2	2.56	7.79	1.70E−17	4,061
RK_30_2321.0B_1	2.50	8.24	2.50E−17	3,762
RK_30_2321.0B_2	2.54	8.47	2.50E−17	4,164
RK_30_2321.2B_1	2.58	7.11	2.38E−17	4,235
RK_30_2321.2B_2	2.55	6.98	2.37E−17	4,122
RK_30_2321.5A	2.57	6.92	2.60E−17	3,965
RK_30_2322.3A_1	2.54	7.61	2.44E−17	4,016
RK_30_2322.3A_2	2.57	6.83	2.27E−17	4,069

Siratovich et al.

Siratovich et al. Geothermal Energy 2014 2:10   doi:10.1186/
s40517-014-0010-4

DISCUSSION

Micromechanical Interpretation

We have shown that the Rotokawa Andesite contains a pervasive network of isotropic microcracks. Due to their isotropic distribution, the majority of these microcracks are consistent with the results of thermal stressing (Fredrich and Wong [1986]; Reuschlé et al. [2006]; Wang et al. [1989]; David et al. [1999]; Heap et al. [2014]). Indeed, the Rotokawa Andesite has experienced several cycles of heating and cooling: the initial eruption of the andesite, burial in a faulted graben, hydrothermal alteration, and the eventual exhumation during core recovery (Rae [2007]; Lim et al. [2012]). Our microstructural analysis has highlighted that the pervasive microcracking appears independent of lithology, original mineralogy, and secondary (hydrothermal alteration) mineralogy.

The intense microcracking in our samples has shown to be a significant factor in all of the measured physical properties. First, microcracking has greatly reduced the propagation velocity of elastic waves through the andesite. We see a clear correlation of crack area per unit volume (Sv) to the observed compressional wave velocities (Figure 8D) and interpret this to be attenuation of the compressional wave through the cracked intracrystalline and intercrystalline boundaries that are abundant in the andesite (e.g., Figures 3 and 4). Many authors (e.g., Vinciguerra et al. [2005]; Keshavarz et al. [2010]; Blake et al. [2012]; Heap et al. [2014]) have also shown that the elastic wave velocities can be highly attenuated by the presence of microcracks.

Second, the crack surface area and UCS have yielded an excellent correlation (Figure 11B). As noted by Walsh ([1965a], [b]), David et al. ([1999]), and Chaki et al. ([2008]), the density of the cracks within a specimen is critical in dictating its strength. The development of microcracks during uniaxial compression, and the coalescence of these cracks (newly formed and pre-existing), leads to the failure of the sample (Brace et al. [1966]; Bieniawski [1967]). In samples that already show relatively high crack densities, less energy is required to coalesce existing cracks and thus they are inherently weaker (David et al. [1999]; Ferrero and Marini [2001]; Keshavarz et al. [2010]). By utilizing AE monitoring during our UCS testing, we observe that fewer events occur during uniaxial compression in weaker samples than those with higher strength (Figure 10), indicating that there are far more pre-existing cracks in the weaker samples (Hardy [1981]; Eberhardt et al.[1998]; Nicksiar and Martin [2012]). Thus, the presence of pre-existing microcracks in the Rotokawa Andesite is shown to exert a strong control on their uniaxial compressive strength.

Permeability is one of the most important properties of a geothermal system. In this study, we have seen that porosity (and bulk sample density) and strength are related to the extent of the microcracking in the andesite. We did not measure the crack surface area in the samples used for our permeability measurements (the samples will be used for future studies; calculating crack surface area required destructive thin section preparation). However, we can, by proxy, assume a correlation between permeability and the extent of the microfracture network. We show that there is a clear inverse relationship between the sample's permeability and P-wave velocity such that as permeability increases, compressional wave velocity decreases (Figure 11F). These results are consistent with the many investigations have shown a clear link between reduced elastic wave velocities and increased permeability (David et al. [1999]; Vinciguerra et al. [2005]; Chaki et al.[2008]; Nara et al. [2011]; Faoro et al. [2013]; Heap et al. [2014]). While we have not measured the relationship of crack density to permeability directly in our dataset, we show that Sv and Vp are inversely related (Figure 8D), and a similar relationship exists between Vp and permeability. Therefore, we can infer that those samples with higher crack surface areas will be inherently more permeable.

Key Empirical Relationships

In this section, we present relationships of singular variables that could be readily and easily measured either using photomicrography or geophysical logging tools and their correlation to more complicated and pertinent physical properties. All of these parameters are singularly measurable variables that do not rely on complex formulae for their derivation (such as dynamic Young's Modulus or Poisson's ratio) and so have been selected to be the key relationships that we present with relevance to the Rotokawa Andesite.

Porosity and UCS

An exponential correlation between sample porosity and UCS exists (Figure 11A). Such correlations have been utilized by several authors (e.g., Vernik et al. [1993]; Li and Aubertin[2003]; Palchik and Hatzor [2002]; Kahraman et al. [2005]; Chang et al. [2006]; Palchik [2013]; Pola et al. [2014]) for a variety of clastic and volcanic rocks and concrete materials. These authors present empirical fits for the correlation of physical properties versus UCS and show a wide range of correlation within their respective datasets with R^2 values from near 0.6 to as high as 0.95. We propose that our empirical fit between porosity and UCS (an exponential fit with a correlation factor of 0.82, Figure 11A) can provide useful estimations of the strength of the reservoir rocks within the Rotokawa Andesite reservoir. By utilizing estimations of UCS derived from the correlation of porosity, the minimum strength of the rocks can be applied to important engineering issues such as wellbore stability (Chang et al. [2006]; Schöpfer et al. [2009]).

Vp and UCS

There is an exponential correlation between strength and Vp with an R^2 value of 0.74 (Figure 11C). As noted by Kahraman ([2001]), the relationship between Vp and UCS is generally nonlinear and the higher the strength of the material, the more scattered the data points. Heap et al. ([2014]) came to similar conclusions following measurements on andesitic rocks from Volcán de Colima (Mexico). In our study, there is an increasing trend of strength with increasing Vp but, as shown in Figure 9, there is a high degree of spatial anisotropy with respect to Vp such that a robust correlation of strength to elastic wave velocity is difficult to obtain. However, Vp is a widely utilized logging tool in borehole geophysics (Chang et al. [2006]), and using the correlation that we have obtained, a minimum strength criteria could be established from the response of the logging tool. This is an important correlation as geophysical logging is much easier, faster, and more efficient than cutting spot cores (as the core for this study was obtained), and so the development of empirical correlations to constrain strength such as that seen in Figure 11B can help mitigate risk and reduce the cost associated with geothermal drilling programs.

Vp and Porosity

Correlations between Vp and porosity show an increasing trend of porosity with decreasing Vp (Figure 11D, also observed by Al-Harthi et al. [1999]; Rajabzadeh et al. [2011]; Tugrul and Gurpinar [1997]; Heap et al. [2014]). This can be attributed to both the pore structure distribution and the degree of microcracking within the andesites. It is clear from microstructural analysis (using both optical and scanning electron microscope analyses) that a large proportion of the porosity in the Rotokawa Andesite is likely to be composed of (macro- and mesoscale) fractures and microcracks (e.g., Figures 6 and 7).

An explanation for the variation and wide distribution of the elastic wave velocity data for samples with similar porosities (specifically with regard to those data that range from 4,000 to 4,400 m/s) is that there must be a variable pore (vug/vesicle) content or hydrothermal alteration between the samples. The presence of pores will greatly augment the porosity (due to their aspect ratio) but will have comparatively little influence, compared to the microcracks, on the P-wave velocity. The application of our exponential relationship (Figure 11D) can give a rough approximation for seismic velocities derived from connected porosity, or *vice versa*. This may be useful during the drilling of additional wells at Rotokawa where porosity can be measured at the wellsite and yield a rough approximation for P-wave velocities and, as such, tie back to our empirical correlations of strength (Figure 11C).

Permeability and Porosity

Our permeability and porosity data show that there is a clear trend of increasing porosity with increased permeability for the Rotokawa

Andesite (Figure 11E), a common observation in multiple lithologies (e.g., Heard and Page [1982]; Géraud [1994]; Stimac et al. [2004]; Chaki et al. [2008]; Watanabe et al. [2008]; Heap et al. [2014]). We observe that our relationship between porosity and permeability can be described by a power law correlation and is consistent with the Kozeny-Carman relation (Guéguen and Palciauskas [1994], see the 'Application of micromechanical and geometrical permeability models' section). The dependence of permeability on porosity is generally explained by the assumption that a more connected pore space (cracks and pores) provides more efficient pathways for fluid migration (e.g., Costa [2006]; Chaki et al. [2008]). We do however need to consider those data points that have a very similar value of permeability (approximately 3.2×10^{-17} m^2, Table 4), with a porosity range of 7.6 to 10.3 vol% that indicate that there is variability of the samples with respect to permeability that may be reflected in the tortuosity of the porous network. This is consistent with the findings of Bernard et al. ([2007]) and Heap et al. ([2014]) such that the permeability in volcanic rocks is highly dependent upon connectivity of the microstructure.

With respect to microstructure, we have shown that the porosity is very closely linked to crack surface area (Figure 8D) and, thus, that increasing crack density corresponds to a sample with a higher permeability. The three samples that lie slightly outside the trend of the dataset display distinct mesofractures (black stars in Figure 11E, F) and that these mesofractures greatly enhance the permeability of the samples without significantly increasing their porosity. These specimens show higher than average permeability for their porosity, which supports the conclusions of Stimac et al. ([2008]) that meso- and macrofractures are critical in controlling the permeability of geothermal reservoir systems. On the large scale, macrofractures are necessary for fluid production from geothermal reservoirs, but the microstructural characteristics of the host rocks cannot be neglected when considering fluid flow, storage capacity, and total permeability of the reservoir (Jafari and Babadagli [2011]).

The robust relationship between porosity and permeability has wider-scale reservoir applications where the need to understand reservoir rock permeability (the mass itself, not those portions with highly macroscopic fractures e.g., Massiot et al. [2012]) is important for reservoir forecasting and modeling. Measurements of porosity can then yield a good approximation of the permeability of the intact reservoir rock at Rotokawa through our power law correlation (Figure 11E). However, we urge caution if the porosity falls outside our measured range. As porosity is a readily measureable property by geophysical logging tools (Ellis and Singer [2008]), the response from such a tool, together with our empirical fit, can give engineers and geoscientists an approximation of the matrix permeabilities in the Rotokawa Andesite.

Permeability and Acoustic Velocities

There is a clear inverse relationship between our measurements of permeability and P-wave velocity (Figure 11F) such that the more permeable the sample, the slower the compressional wave velocity. These findings are consistent with the findings of many other authors (e.g., Vinciguerra et al. [2005]; Chaki et al. [2008]; Nara et al. [2011]; Heap et al. [2014]). The correlation of such properties is an excellent tool for understanding the micro- and mesoscopic fracture networks and their relation to permeability in the Rotokawa Andesite as follows: (1) we have shown that the porosity and crack density are closely linked (Figure 8A), (2) acoustic velocity and crack density are closely linked (Figure 8D), and (3) there is a power law correlation of Vp and permeability (Figure 11F). Thus, there is a direct link of P-wave velocity to permeability that is reliant on the crack densities of the samples. The relationship we present in Figure 11F shows a power-law fit which would indicate that the hydraulic radii of the pore space (pore and cracks) are similar in size but that the higher the concentration of cracks, the higher the permeability we observe (Bourbie and Zinszner [1985]).

Similarly, there are occasional mesofractures (with apertures less than 1-mm width; we note that these fractures are much smaller than those described in Massiot et al. [2012]) in the samples that deviate from the rest of the dataset (black stars, Figure 11F). The presence of these macrofractures increases permeability (by a factor of 2) and also appears deleterious to elastic wave propagation (all the three samples containing mesofractures have low elastic wave velocities, although we cannot separate the influence of meso- and microcracks on the velocities of these samples). Further, elastic waves are useful for the detection of cracks in rock and concrete (Chaki et al. [2008]; Heap et al. [2013]), and a decreased elastic wave velocity correlates well to more permeable media which is observed by the three outlying, higher permeability, lower elastic wave velocity samples.

The correlation between elastic wave velocity and permeability outside the laboratory has potentially far-reaching value for the prediction of reservoir permeability interactions from wireline logging and larger-scale seismic and microseismic surveys. There is a complex microseismic network installed at Rotokawa, and the location of earthquake activity has been closely linked to macroscopic permeability within the reservoir (Sewell et al. [2013]; Sherburn et al. [2013]). The existing model of the velocity structure at depth could then be further refined using our acoustic velocity and permeability data for reservoir rock matrix. This may allow a deeper and more accurate understanding of the distribution of permeability at depth.

Additionally, the data we have presented can also be used to infer values of matrix permeability from acoustic wireline logs (dipole sonic) used during exploration at nearby Ngatamariki Geothermal Field (Wallis et al. [2009]). Should similar geophysical logging be used in future wells drilled at Rotokawa, the matrix permeability may be estimated using the relationship we present here. In addition, the coupling of these data with microseismic data could allow a significant increase in understanding the complexity of the Rotokawa Andesite reservoir. While we are aware that macrofractures augment the elastic wave velocity during routine acoustic profiling (e.g., Barton and Zoback [1992]), our laboratory data show that although samples containing mesofractures (i.e., on the sample scale) are shifted to higher permeabilities and elastic wave velocities, they do not stray too far away from the trend extrapolated from our power-law relationship. Despite this, we urge a certain degree of caution, based on the potential presence of large-scale fractures, when estimating permeability using our derived permeability-elastic wave velocity relationship.

Application of Micromechanical and Geometrical Permeability Models

Extracting empirical relationships between laboratory-derived rock properties is useful; however, the parameters are not easily related to independently measurable quantities (i.e., they lack a physical basis). Micromechanical (e.g., the wing-crack model of Ashby and Sammis [1990]) and geometrical permeability models (e.g., the Kozeny-Carman relation, Guéguen and Palciauskas[1994]) can be better constrained as the parameters used in such models have a clear physical meaning. In this section, we attempt both sliding wing-crack modeling and Kozeny-Carman permeability modeling to investigate the microstructural controls on deformation and fluid flow, respectively.

Micromechanical Modeling

Micromechanical modeling can provide useful insights in the mechanics of compressive failure in brittle rock (Wong and Baud [2012]). Since the rocks of this study contain high microcrack densities, we will use the sliding wing-crack model of Ashby and Sammis ([1990]). This model idealizes the rock microstructure as an elastic continuum embedded with inclined (45°) microcracks (of length 2c). These microcracks act as stress concentrators for the initiation of 'wing' cracks when the frictional resistance of the closed crack is overcome and the stress at the tip of the crack exceeds the critical stress intensity factor (K_{IC}). The cracks can then propagate in the direction of the maximum principal stress. Eventually, the cracks coalesce, resulting in the failure of the elastic medium. In the case of uniaxial compression, Baud et al. ([2014]) derived an analytical approximation to estimate UCS:

$$UCS = \frac{1.346}{\sqrt{1+\mu^2}-\mu} \frac{K_{Ic}}{\sqrt{\pi c}} D_0^{-0.256}$$

(10)

where μ is the friction coefficient of the sliding crack and D_0 is an initial damage parameter that is a function of the angle of the initial microcrack with respect to the maximum principal stress and the initial number of sliding cracks per unit area (Ashby and Sammis [1990]).

The analytical solution (that assumes an initial crack angle of 45°) presented above contains five parameters. We have, through experimental data and observations, a good handle on three of the parameters: (1) we have measured the UCS of 22 samples (Table 3), (2) μ rarely deviates from 0.6 to 0.7 (Byerlee [1978]), and (3) c can be determined from optical microscopy (we determinedc by measuring the approximate average length of the microcracks under the microscope). We do not have a laboratory-determined value for K_{IC}. While the K_{IC} of andesite has been previously measured to be about 1.5 MPam$^{0.5}$ (Ouchterlony [1990]; Obara et al. [1992]; Tutluoglu and Keles[2011]; Nara et al. [2012]), there is no guarantee that this value is representative of the Rotokawa Andesite, which is likely to be lower than these values due to hydrothermal alteration. We therefore have chosen a slightly lower K_{IC} of 1.0 MPam$^{0.5}$ for our analysis. Using our UCS data, we can solve Equation 10 to assign a value of D_0 to each experiment (using $\mu = 0.6$; $K_{IC} = 1.0$; $c = 0.001$ m). The goal of such analysis, assuming that the other parameters remain roughly constant between different samples/cores, is to estimate D_0 using an easily measured physical property, such as Vp (therefore allowing us to predict rock strength, using the micromechanical model, from Vp measurements alone). Our analysis shows that D_0 ranges from 0.0019 to 0.26 for the 22 measured samples (with average of 0.039). D_0 is plotted against the crack area per unit volume (Sv) and Vp in Figure 12 and indicates that D_0 increases as Sv increases (Figure 12A). While this may appear logical (D_0 is a function of the initial crack density), it serves as an encouraging proof of the concept. The increase in D_0 with crack density is not linear; D_0 increases more rapidly beyond 10 mm^{-1} (Figure 12A). We also see that Vp decreases with increasing D_0; in detail, Vp decreases rapidly as D_0 increases from 0 to 0.05 and then decreases more gradually above 0.05. Unfortunately, the relationship between D_0 and Vp is a little more clouded (the data are more scattered, Figure 12B) and probably represents variable vesicle density (the model assumes that vesicles do not play a role in failure in compression) and hydrothermal alteration (we assume that K_{IC} and the average crack lengths are constant). The conclusion of this pilot analysis is that the variability within the Rotokawa Andesite is potentially too large to permit meaningful microstructural wing-crack modeling, but greater success could be achieved with laboratory-determined values for K_{IC}. Therefore, if micromechanical modeling is to be deployed as a feasible method to predict the strength of Rotokawa Andesite reservoir rocks, the samples/cores should be grouped by their alteration, and K_{IC} measured for each alteration group.

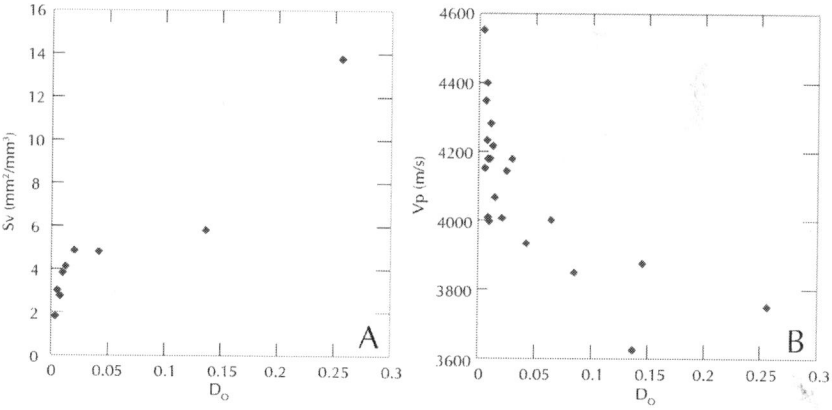

Figure 12: Results of geometric modeling for Rotokawa Andesite. (A) Initial damage parameter D_0 as predicted by Equation 10 and described by Baud et al. ([2014]) plotted versus calculated crack densities by the method of (Underwood [1970]). (B) Prediction of compressional wave velocity (Vp) as a function of the initial damage parameter D_0 the relationship between D_0 and Vp shows a moderate correlation with high initial damage

parameter but becomes quite clouded in those samples with a very small calculated D_0 (see text for further expansion on this relationship).

Permeability Modeling

Kozeny-Carman models are those that use the notion of a hydraulic radius (see Guéguen and Palciauskas [1994]) to correlate porosity and permeability. Forms of the Kozeny-Carman relation have previously been used in the study of volcanic rocks (e.g., Saar and Manga [1999]; Costa[2006]; Bernard et al. [2007]; Heap et al. [2014]), while others have used a heavily simplified version (e.g., Rust et al. [2003]; Mueller et al. [2005]; Lavallée et al. [2013]). The Kozeny-Carman relation is of the form:

$$k_{\text{KC}} = \frac{\varphi(r_{\text{H}})^2}{b\tau^2}$$

(11)

where k_{KC} is the permeability, φ is the connected porosity, b is a geometrical factor, τ is the tortuosity of the equivalent channel (i.e., the ratio of its actual to nominal length), and r_{H} is the hydraulic radius (i.e., the volume of pores divided by the surface of the pores). The power law exponent for our data (excluding those samples with macrofractures) is about 2.2 (Figure 11E) and is therefore consistent with the Kozeny-Carman model (Bourbie and Zinszner [1985]; Doyen[1988]). In detail, one would expect a power law exponent of 2 or 3 if the elements controlling the permeability are tubes or cracks, respectively (Guéguen and Palciauskas [1994]). Our power law exponent is between these two values. This is somewhat surprising, considering the pervasive fracture network in these materials, but could reflect flow through a combination of cracks and tubes or our limited porosity range. Since the entire dataset can be described by a single power law exponent, we conclude that within our limited range of connected porosities, there is no dramatic shift in pore space connectivity or tortuosity, as was the case for Fontainebleau sandstone at a porosity of 9 vol% (Bourbie and Zinszner [1985]) and andesite samples from Volcán de Colima (Mexico) at a porosity of about 11 vol% (Heap et al. [2014]).

Extrapolating to porosities outside this range may be treacherous especially to lower porosities where samples may become subject to a higher power law exponent. However, within the dataset, the model predicts an increase in permeability of a factor of 1.5 for an increase in porosity of 1 vol% (an increase not uncommon for rock following a thermal stressing episode; e.g., Chaki et al. [2008]).

Application of Results to Geothermal Exploration and Utilization

The relationships between porosity, acoustic wave velocities, strength, and permeability are valuable for understanding a geothermal reservoir. Our data indicate strong correlations between these parameters, as observed by Stimac et al. ([2004], [2008]) amongst others. The data we have obtained are from cores sourced from three production wells. Such materials are very expensive to obtain, time consuming, and, if coring did not go as planned, can pose great risk of losing the well (Finger and Blankenship [2010]; Hole [2013]). The microstructural and empirical correlations presented in this study can be applied to new wells drilled in geothermal environments and can help refine studies on pre-existing wells, if our correlations hold true at the reservoir scale. Some physical parameters, such as porosity and elastic wave velocities, are easily obtainable through the use of down-hole geophysical logging suites. The empirical correlations shown in this study (bolstered by our application of classical models) show that readily measurable physical properties may therefore be used to predict more complex and pertinent properties such as strength and permeability. Such correlations and calibrations are common in the hydrocarbon industry especially during exploration drilling (e.g., Vernik et al. [1993] and references therein), and we consider that our dataset can help improve the understanding of the Rotokawa reservoir while minimizing the risk to future drilling operations.

A clear understanding of the factors that control reservoir rock permeability is fundamental for the planning of stimulation and enhancement operations that may be necessary as the Rotokawa field and reservoir dynamics change with continued production. The need to drill additional wells or re-work pre-existing wells may become apparent and the ease at which the reservoir can accept and deliver fluids (i.e., its permeability) will be of utmost importance. The thermal stimulation of injection wells has taken place at Rotokawa for some time by the injection of power-plant condensates and spent brines (Siega et al. [2009]), but the technique may play a significant role in enhancing production wells at some future stage.

Therefore, a deeper understanding of how permeability may be increased through stimulation is important. The application of models such as the Kozeny-Carman may provide insight to permeability enhancement. An increase in the porosity of reservoir rock by 1 vol%, according to the geometrical model, should increase the permeability by a factor of 1.5. In the case of an aging field and aging wellbores, such an increase could greatly extend the life of the field. In the interests of keeping geothermal projects commercially economic, the fundamental understanding of the reservoir rock properties become essential to the continued utilization and management of the field.

CONCLUSIONS

Our study provides a comprehensive evaluation of the physical and mechanical properties of the Rotokawa Andesite through a multi-disciplinary approach. We have evaluated the Rotokawa Andesite from the microstructural to macroscopic scale and have presented robust datasets that permit the correlation and comparison of important physical properties to geothermal exploitation. A comprehensive understanding of how the relationships of microstructural texture influence key physical properties such as strength and permeability, essential for the optimal utilization of a geothermal resource have been investigated.

Further, we summarize our conclusions as follows:

- We have shown that the presence and intensity of microfracturing in the Rotokawa Andesite are the predominant controlling factors on physical and mechanical properties. The behavior of these properties is also shown to be largely independent of the alteration mineralogy as we see similar alteration intensities in the samples we have studied.

- Guided by a systematic understanding of role of microfractures, we show that empirical correlations of strength and porosity can be developed and applied to field scale engineering problems. We have shown that as the porosity increases, the strength decreases and elastic wave velocities are attenuated. Similarly, we show that permeability increases with increased porosity and reduced acoustic velocity. These findings are applicable if geophysical logging tools be used after the drilling of wells to ascertain properties such as porosity; our dataset provides useful means to address complex reservoir problems.

- We further boost our empirical correlations by applying classical physical models based on sound physical theory to predict both UCS and permeability through understanding of the microstructure. We have applied these models with some success, but these models are best-suited for homogeneous, isotropic materials. Further work to constrain these models should include laboratory investigations of fracture toughness (K_{IC}) and the factors that influence this variable. However, our fit for the damage criterion D_0 is acceptable and builds the foundations for future understanding and may permit the construction of similar better constrained models.

- The study comprises a large dataset with a goal to further push the knowledge that can be sourced from a geothermal environment such as the Rotokawa Andesite. The properties that we have evaluated are very difficult to constrain without direct information from rocks sourced from the reservoir. Geothermal reservoirs are complex, and harsh environments from which the recovery of intact core can present a significant and financially risky challenge. The results that we present here help us to understand this complex reservoir environment by their application to field scale engineering and geological issues.

- Our analyses have provided quantifiable and measurable physical properties of the Rotokawa Andesite. However, the dataset is not exhaustive. Further studies need to be carried out to replicate near-reservoir conditions in the laboratory and should focus on permeability at the high confining pressures and temperatures found in the reservoir. Additionally, mechanical testing such as triaxial, tensile strength, and fracture toughness experiments should be conducted under high-temperature conditions, potentially in the presence of reservoir-type fluids to aid in predictions of reservoir behavior and geomechanical modeling under conditions as close as possible to those found in the reservoir.

AUTHORS' CONTRIBUTIONS

PS performed the majority of the experimental work and composition of the manuscript. MH assisted in revision and suggestions for the manuscript and performed the permeability and micromechanical modeling. MV provided assistance in experimental work and processing of mechanical data. JC provided revision and petrological input. TR provided the permeability measurements. All authors read and approved the final manuscript.

ACKNOWLEDGMENTS

The authors wish to thank Mighty River Power Company Ltd. for a generous grant for PS, which allowed collaboration with MH and TR. We also wish to thank the Rotokawa Joint Venture, a joint venture between the Tauhara North No. 2 Trust and Mighty River Power Company Ltd. for the core material used in this study. The staff of the Department of Geological Sciences at the University of Canterbury were invaluable in assisting in all aspects of this research. The Brian Mason Trust also provided for the transportation and delivery of the core to UC. The authors of this study also acknowledge a Hubert Curien Partnership (PHC) Dumont d'Urville grant (grant number 31950RK) which has assisted the France-New Zealand collaboration for this and future projects. MH was partly funded by the LABEX ANR-11-LABX-0050_G-EAU-THERMIE-PROFONDE framework (funding from the state managed by the French National Research Agency as part of the Investments for the Future Program).

REFERENCES

1. Al-Harthi AA, Al-Amri RM, Shehata WM (1999) The porosity and engineering properties of vesicular basalt in Saudi Arabia. Eng Geol 54:313-320

2. Ashby MF, Sammis CG (1990) The damage mechanics of brittle solids in compression. Pure Appl Geophys 133:489-521

3. Barton CA, Zoback MD (1992) Self-similar distribution and properties of macroscopic fractures at depth in crystalline rock in the Cajon Pass Scientific Drill Hole. J Geophys Res 97(B4):5181-5200

4. Baud P, Wong T-F, Zhu W (2014) Effects of porosity and crack density on the compressive strength of rocks. Int J Rock Mech Min Sci   :   (in press)

5. Bernard M-L, Zamora M, Géraud Y, Boudon G (2007) Transport properties of pyroclastic rocks from Montagne Pelée volcano (Martinique, Lesser Antilles). J Geophys Res 112:

6. Bibby HM, Caldwell TG, Davey F, Webb T (1995) Geophysical evidence on the structure of the Taupo Volcanic Zone and its hydrothermal circulation. J Volcanol Geotherm Res 68:29-58

7. Bieniawski ZT (1967) Mechanism of brittle fracture of rock part II - experimental studies. Int J Rock Mech Min Sci 4:407-423

8.   Blake OO, Faulkner DR, Rietbrock A (2012) The effect of varying damage history in crystalline rocks on the P- and S-wave velocity under hydrostatic confining pressure. Pure Appl Geophys 170:493-505

9.   Bloomberg S, Rissmann C, Mazot A, Oze C, Horton T, Kennedy B, Werner C, Christenson B, Pawson J (2012) Soil gas flux exploration at the Rotokawa Geothermal Field and White Island, New Zealand. In: Proceedings, Thirty Sixth Workshop on Geothermal Reservoir Engineering. Stanford University, Stanford, California. 30 January 30 to 1 February 2012

10.   Bourbie T, Zinszner B (1985) Hydraulic and acoustic properties as a function of porosity in Fontainebleau sandstone. J Geophys Res 90:11524-11532

11.   Brace WF, Bombolakis EG (1963) A note on brittle crack growth in compression. J Geophys Res 68:3709-3713

12.   Brace WF, Paulding B, Scholz C (1966) Dilatancy in the fracture of crystalline rocks. J Geophys Res 71:3939-3953

13.   Brace WF, Walsh JB, Frangos WT (1968) Permeability of granite under high pressure. J Geophys Res 73:2225-2236

14.   Byerlee JD (1978) Friction of rocks. Pure Appl Geophys 116:615-626

15.   Chaki S, Takarli M, Agbodjan WP (2008) Influence of thermal damage on physical properties of a granite rock: porosity, permeability and ultrasonic wave evolutions. Constr Build Mater 22:1456-1461

16.   Chang C, Zoback MD, Khaksar A (2006) Empirical relations between rock strength and physical properties in sedimentary rocks. J Pet Sci Eng 51:223-237

17.   Cole JW (1990) Structural control and origin of volcanism in the Taupo volcanic zone, New Zealand. Bull Volcanol 52:445-459

18.   Collar RJ, Browne PRL (1985) Hydrothermal eruptions at The Rotokawa Geothermal Field, Taupo Volcanic Zone, New Zealand. In: Proceedings of the seventh New Zealand

geothermal workshop, University of Auckland. Geothermal Institute, Auckland, New Zealand. 6–8 November 1985

19. Costa A (2006) Permeability-porosity relationship: a re-examination of the Kozeny-Carman equation based on a fractal pore-space geometry assumption. Geophys Res Lett 33:

20. David C, Menendez B, Darot M (1999) Influence of stress-induced and thermal cracking on physical properties and microstructure of La Peyratte granite. Int J Rock Mech Min Sci 36:433-448

21. Diamantis K, Gartzos E, Migiros G (2009) Study on uniaxial compressive strength, point load strength index, dynamic and physical properties of serpentinites from Central Greece: test results and empirical relations. Eng Geol 108:199-207

22. Diederichs M, Kaiser P, Eberhardt E (2004) Damage initiation and propagation in hard rock during tunneling and the influence of near-face stress rotation. Int J Rock Mech Min Sci 41:785-812

23. DiPippo R (2008) Geothermal power plants: principles, applications, case studies and environmental impact. Elsevier Ltd, Oxford.

24. Doyen PM (1988) Permeability, conductivity, and pore geometry of sandstone. J Geophys Res 93:7729-7740

25. Eberhardt E, Stead D, Stimpson B, Read RS (1998) Identifying crack initiation and propagation thresholds in brittle rock. Can Geotech J 35:222-233

26. Ellis DV, Singer JM (2008) Well logging for earth scientists. Springer, Dordrecht.

27. Faoro I, Vinciguerra S, Marone C, Elsworth D, Schubnel A (2013) Linking permeability to crack density evolution in thermally stressed rocks under cyclic loading. Geophys Res Lett 40:2590-2595

28. Ferrero AM, Marini P (2001) Experimental studies on the mechanical behaviour of two thermal cracked marbles. Rock Mech Rock Eng 34:57-66

29. Finger J, Blankenship D (2010) Handbook of best practices for geothermal drilling. Sandia National Laboratories, Albuquerque.

30. Fredrich JT, Wong T (1986) Micromechanics of thermally induced cracking in three crustal rocks. J Geophys Res 91:12743-12764

31. Géraud Y (1994) Variations of connected porosity and inferred permeability in a thermally cracked granite. Geophys Res Lett 21:979-982

32. Grant MA, Bixley PF (2011) Geothermal reservoir engineering. Elsevier Science Ltd, Oxford.

33. Guéguen Y, Palciauskas V (1994) Introduction to the physics of rocks. Princeton University Press, Princeton.

34. Guéguen Y, Schubnel A (2003) Elastic wave velocities and permeability of cracked rocks. Tectonophysics 370:163-176

35. Gupta H, Sukanta R (2006) Geothermal energy: an alternative resource for the 21st century. Elsevier B.V., Oxford.

36. Hardy H (1981) Applications of acoustic emission techniques to rock and rock structures: a state of the art review. In: Drnevich V, Gray R (eds) Acoustic emission in geotechnical engineering practice, American Society for Testing and Materials, University of Michigan, Ann Arbor.

37. Heap MJ, Faulkner DR (2008) Quantifying the evolution of static elastic properties as crystalline rock approaches failure. Int J Rock Mech Min Sci 45:564-573

38. Heap MJ, Vinciguerra S, Meredith PG (2009) The evolution of elastic moduli with increasing crack damage during cyclic stressing of a basalt from Mt. Etna volcano. Tectonophysics 471:153-160

39. Heap MJ, Lavallée Y, Laumann A, Hess K-U, Meredith PG, Dingwell DB, Huismann S, Weise F (2013) The influence of thermal-stressing (up to 1000°C) on the physical, mechanical, and chemical properties of siliceous-aggregate, high-strength concrete. Construct Build Mater 42:248-265

40. Heap MJ, Lavallee Y, Petrakova L, Baud P, Reushcle T, Varley NR, Dingwell DB (2014) Microstructural controls on the physical and mechanical properties of edifice-forming andesites at Volcán de Colima Mexico. J Geophys Res 119:2925-2963

41. Heard HC, Page L (1982) Elastic moduli, thermal expansion, and inferred permeability of two granites to 350°C and 55 megapascals. J Geophys Res 87:9340-9348

42. Hedenquist JW, Mroczek EK, Giggenbach WF (1988) Geochemistry of the Rotokawa geothermal system: summary of data, interpretation and appraisal for energy development. In: Chemistry Division DSIR Technical Note 88/6.   ,   .

43. Hole HM (2013) Geothermal drilling - keep it simple. In: Proceedings of the 35th New Zealand geothermal workshop. , Rotorua, New Zealand. 17–20 November 2013

44. Horie T, Muto T (2010) The world's largest single cylinder geothermal power generation unit - Nga Awa Purua Geothermal Power Station, New Zealand. Geothermal Res Council Trans 34:1039-1044

45. Jafari A, Babadagli T (2011) Effective fracture network permeability of geothermal reservoirs. Geothermics 40:25-38

46. Jaya MS, Shapiro SA, Kristinsdóttir LH, Bruhn D, Milsch H, Spangenberg E (2010) Temperature dependence of seismic properties in geothermal rocks at reservoir conditions. Geothermics 39:115-123

47. Ju Y, Yang Y, Peng R, Mao L (2013) Effects of pore structures on static mechanical properties of sandstone. J Geotech Geoenvironmental Eng 139:1745-1755

48. Kahraman S (2001) Evaluation of simple methods for assessing the uniaxial compressive strength of rock. Int J Rock Mech Min Sci 38:981-994

49. Kahraman S, Gunaydin O, Fener M (2005) The effect of porosity on the relation between uniaxial compressive strength and point load index. Int J Rock Mech Min Sci 42:584-589

50. Keshavarz M, Pellet FL, Loret B (2010) Damage and changes in mechanical properties of a gabbro thermally loaded up to 1,000°C. Pure Appl Geophys 167:1511-1523

51. Klinkenberg LJ (1941) The permeability of porous media to liquids and gases. Drilling and production practice. American Petroleum Institute, New York.

52. Kristinsdóttir LH, Flóvenz ÓG, Árnason K, Bruhn D, Milsch H, Spangenberg E, Kulenkampff J (2010) Electrical conductivity and P-wave velocity in rock samples from high-temperature Icelandic geothermal fields. Geothermics 39:94-105

53. Krupp RE, Seward TM (1987) The Rotokawa geothermal system, New Zealand; an active epithermal gold-depositing environment. Econ Geol 82:1109-1129

54. Lavallée Y, Benson PM, Heap MJ, Hess KU, Flaws A, Schillinger B, Meredith PG, Dingwell DB (2013) Reconstructing magma failure and the degassing network of dome-building eruptions. Geology 41:515-518

55. Legmann H, Sullivan P (2003) The 30 MW Rotokawa I geothermal project five years of operation. In: International geothermal conference.   , Reykjavik, Iceland. September 2003

56. Li L, Aubertin M (2003) A general relationship between porosity and uniaxial strength of engineering materials. Can J Civ Eng 30:644-658

57. Lim SS, Martin CD, Åkesson U (2012) In-situ stress and microcracking in granite cores with depth. Eng Geol 147–148:1-13

58. Lion M, Skoczylas F, Ledésert B (2005) Effects of heating on the hydraulic and poroelastic properties of Bourgogne limestone. Int J Rock Mech Min Sci 42:508-520

59. Luping T (1986) A study of the quantitative relationship between strength and pore-size distribution of porous materials. Cem Concr Res 16:87-96

60. Lutz SJ, Hickman S, Davatzes N, Zemach E, Drakos P, Robertson-Tait A (2010) Rock mechanical testing and

petrologic analysis in support of well stimulation activities at the Desert Peak Geothermal Field, Nevada. In: Proceedings of the thirty-fifth workshop on geothermal reservoir engineering. Stanford University, Stanford, California. 1–3 February 2010

61. Martin CD (1993) The strength of Massive Lac du Bonnet granite around underground openings. Dissertation, University of Manitoba.

62. Martin CD, Chandler NA (1994) The progressive fracture of Lac du Bonnet granite. Int J Rock Mech Min Sci 31:643-659

63. Martínez-Martínez J, Benavente D, García-del-Cura MA (2011) Spatial attenuation: the most sensitive ultrasonic parameter for detecting petrographic features and decay processes in carbonate rocks. Eng Geol 119:84-95

64. Massiot C, McNamara D, Lewis B, Price L, Bignall G (2012) Statistical corrections of fracture sampling bias in boreholes from acoustic televiewer logs. In: New Zealand geothermal workshop proceedings.   , Auckland, New Zealand. 19–21 November 2012

65. Mueller S, Melnik O, Spieler O (2005) Permeability and degassing of dome lavas undergoing rapid decompression: an experimental determination. Bull Volcanol 67:526-538

66. Nara Y, Meredith PG, Yoneda T, Kaneko K (2011) Influence of macro-fractures and micro-fractures on permeability and elastic wave velocities in basalt at elevated pressure. Tectonophysics 503:52-59

67. Nara Y, Morimoto K, Hiroyoshi N, Yoneda T, Kaneko K, Benson PM (2012) Influence of relative humidity on fracture toughness of rock: implications for subcritical crack growth. Int J Solids Struct 49:2471-2481

68. Nicksiar M, Martin CD (2012) Evaluation of methods for determining crack initiation in compression tests on low-porosity rocks. Rock Mech Rock Eng 45:607-617

69. Obara Y, Sakaguchi K, Nakayama T, Sugawara K (1992) Anisotropy effect on fracture toughness of rocks. Int J Rock Mech Min Sci Geomech 30:137.

70. Ouchterlony F (1990) Fracture toughness testing of rock with core based specimens. Eng Fract Mech 35:351-366

71. Palchik V (2013) Is there a link between the type of the volumetric strain curve and elastic constants, porosity, stress and strain characteristics? Rock Mech Rock Eng 46:315-326

72. Palchik V, Hatzor YH (2002) Crack damage stress as a composite function of porosity and elastic matrix stiffness in dolomites and limestones. Eng Geol 63:233-245

73. Pereira J-M, Arson C (2013) Retention and permeability properties of damaged porous rocks. Comput Geotech 48:272-282

74. Pola A, Crosta G, Fusi N, Barberini V, Norini G (2012) Influence of alteration on physical properties of volcanic rocks. Tectonophysics 566–567:67-86

75. Pola A, Crosta GB, Fusi N, Castellanza R (2014) General characterization of the mechanical behaviour of different volcanic rocks with respect to alteration. Eng Geol 169:1-13

76. Powell T (2011) Natural subsidence at the Rotokawa Geothermal Field and implications for permeability development. Geothermal Res Council Trans 35:973-976

77. Quinao J, Sirad-Azwar L, Clearwater J, Hoepfinger V, Le Brun M, Bardsley C (2013) Analyses and modeling of reservoir pressure changes to interpret the Rotokawa Geothermal Field response to Nga Awa Purua Power Station operation. In: Proceedings of the 38th workshop on geothermal reservoir engineering. Stanford University, Stanford, California.11–13 February 2013

78. Rae A (2007) Rotokawa geology and geophysics. GNS Science consultancy report 2007/83 May 2007. GNS Science, Lower Hutt.

79. Rae AJ, McCoy-West AJ, Ramirez LE, Alcaraz SA (2009) Geology of production well RK28. Rotokawa Geothermal Field. GNS Science consultancy report 2009/253 September 2009. GNS Science, Lower Hutt.

80. Rae AJ, McCoy-West AJ, Ramirez LE, McNamara D (2010) Geology of production wells RK30L1 and RK30L2, Rotokawa Geothermal Field. GNS Science consultancy report 2010/02 January 2010. GNS Science, Lower Hutt.

81. Rajabzadeh MA, Moosavinasab Z, Rakhshandehroo G (2011) Effects of rock classes and porosity on the relation between uniaxial compressive strength and some rock properties for carbonate rocks. Rock Mech Rock Eng 45:113-122

82. Ramirez LE, Hitchcock D (2010) Geology of production well RK27L2, Rotokawa Geothermal Field, GNS Science Consultancy Report 2010/100 April 2010. GNS Science, Lower Hutt.

83. Reuschlé T, Gbaguidi Haore S, Darot M (2006) The effect of heating on the microstructural evolution of La Peyratte granite deduced from acoustic velocity measurements. Earth Planet Sci Lett 243:692-700

84. Richter D, Simmons G (1977) Microcracks in crustal igneous rocks: microscopy. In: Heacock JG, Keller GV, Oliver JE, Simmons G (eds) The earth's crust, American Geophysical Union, Washington, DC.

85. Rowland JV, Sibson RH (2004) Structural controls on hydrothermal flow in a segmented rift system, Taupo Volcanic Zone, New Zealand. Geofluids 4:259-283

86. Rowland JV, Wilson CJN, Gravley DM (2010) Spatial and temporal variations in magma-assisted rifting, Taupo Volcanic Zone, New Zealand. J Volcanol Geotherm Res 190:89-108

87. Rust AC, Manga M, Cashman KV (2003) Determining flow type, shear rate and shear stress in magmas from bubble shapes and orientations. J Volcanol Geotherm Res 122:111-132

88. Rutter EH (1986) On the nomenclature of mode of failure transitions in rocks. Tectonophysics 122:381-387

89. Saar MO, Manga M (1999) Permeability-porosity relationship in vesicular basalts. Geophys Res Lett 26:111-114

90. Schöpfer MPJ, Abe S, Childs C, Walsh JJ (2009) The impact of porosity and crack density on the elasticity, strength and friction of cohesive granular materials: insights from DEM modelling. Int J Rock Mech Min Sci 46:250-261

91. Sewell SM, Cumming WB, Azwar L, Bardsley C (2012) Integrated MT and natural state temperature interpretation for a conceptual model supporting reservoir numerical modelling and well targeting at the Rotokawa Geothermal Field, New Zealand. In: Proceedings of the thirty-seventh workshop on geothermal reservoir engineering. Stanford University, Stanford California. 30 January to 1 February 2012

92. Sewell SM, Cumming W, Bardsley CJ, Winick J, Quinao J, Wallis IC, Sherburn S, Bourguignon S, Bannister S (2013) Interpretation of microearthquakes at the Rotokawa Geothermal Field, 2008 to 2012. In: Proceedings of the 35th New Zealand geothermal workshop.   , Rotorua, New Zealand. 17–20 November 2013

93. Sherburn S, Bourguignon S, Bannister S, Sewell S, Cumming B, Bardsley C, Quinao J, Wallis I (2013) Microseismicity at Rotokawa Geothermal Field, 2008 to 2012. In: Proceedings of the 35th New Zealand geothermal workshop.   , Rotorua, New Zealand. 17–20 November 2013

94. Siega C, Grant M, Powell T (2009) Enhancing injection well performance by cold water stimulation in Rotokawa and Kawerau geothermal field. In: Proceedings of PNOC-EDC conference.   , Manila, Philippines. 27–28 September 2009

95. Smith R, Sammonds PR, Kilburn CRJ (2009) Fracturing of volcanic systems: experimental insights into pre-eruptive conditions. Earth Planet Sci Lett 280:211-219

96. Sousa LMO, Suárez del Río LM, Calleja L, Ruiz de Argandoña VG, Rey AR (2005) Influence of microfractures and porosity on the physico-mechanical properties and weathering of ornamental granites. Eng Geol 77:153-168

97. Stanchits S, Vinciguerra S, Dresen G (2006) Ultrasonic velocities, acoustic emission characteristics and crack damage

of basalt and granite. Pure Appl Geophys 163:975-994

98. Stimac JA, Powell TS, Golla GU (2004) Porosity and permeability of the Tiwi geothermal field, Philippines, based on continuous and spot core measurements. Geothermics 33:87-107

99. Stimac J, Nordquist G, Suminar A, Sirad-Azwar L (2008) An overview of the Awibengkok geothermal system, Indonesia. Geothermics 37:300-331

100. Takarli M, Prince W, Siddique R (2008) Damage in granite under heating/cooling cycles and water freeze thaw condition. Int J Rock Mech Min Sci 45:1164-1175

101. Tugrul A, Gurpinar O (1997) The effect of chemical weathering on the engineering properties of Eocene basalts in northeastern Turkey. Environ Eng Geosci 3:225-234

102. Tutluoglu L, Keles C (2011) Mode I fracture toughness determination with straight notched disk bending method. Int J Rock Mech Min Sci 48:1248-1261

103. Ulusay R, Hudson J (2007) The complete ISRM suggested methods for rock characterization, testing and monitoring: 1974–2006. Elsevier, Antalya, Turkey.

104. Underwood EE (1970) Quantitative stereology for microstructural analysis. In: Underwood EE (ed) Quantitative stereology, Addison-Wesley, Reading, Massachusetts.

105. Vernik L, Bruno M, Bovberg C (1993) Empirical relations between compressive strength and porosity of siliciclastic rocks. Int J Rock Mech Min Sci Geomech 30:677-680 Abstract

106. Vinciguerra S, Trovato C, Meredith P, Benson P (2005) Relating seismic velocities, thermal cracking and permeability in Mt. Etna and Iceland basalts. Int J Rock Mech Min Sci 42:900-910 Wallis I, McCormick S, Sewell S, Boseley C (2009) Formation assessment in geothermal using wireline tools - application and early results from the Ngatamariki Geothermal Field, New Zealand. In: Proceedings of the New Zealand Geothermal Workshop.   , Rotorua, New Zealand.16–18 November 2009

107. Walsh JB (1965) The effect of cracks on the compressibility of rock. J Geophys Res 70:381-389

108. Walsh JB (1965) The effect of cracks in rocks on Poisson's ratio. J Geophys Res 70:5249-5257

109. Wang HF, Bonner BP, Carlson SR, Kowallis BJ, Heard HC (1989) Thermal stress cracking in granite. J Geophys Res 94:1745-1758

110. Watanabe T, Shimizu Y, Noguchi S, Nakada S (2008) Permeability measurements on rock samples from Unzen Scientific Drilling Project Drill Hole 4 (USDP-4). J Volcanol Geotherm Res 175:82-90

111. Wilson CJN, Houghton BF, Mcwilliams MO, Lanphere MA, Weaver SD, Briggs RM (1995) Volcanic and structural evolution of Taupo Volcanic Zone, New Zealand: a review. J Volcanol Geotherm Res 68:1-28

112. Wong TF, Baud P (2012) The brittle-ductile transition in porous rock: a review. J Struct Geol 44:25-53.

113. Wu XY, Baud P, Wong TF (2000) Micromechanics of compressive failure and spatial evolution of anisotropic damage in Darley Dale sandstone. Int J Rock Mech Min Sci 37:143-160.

Chapter 2

Rheological Study of Partially Hydrolyzed Polyacrylamide-hexamine-pyrocatechol Gel System

Upendra Singh Yadav and Vikas Mahto

Department of Petroleum Engineering, Indian School of Mines, Dhanbad Jharkhand, 826004, India

ABSTRACT

Background

The cross-linked polymer gels exhibit non-Newtonian rheological behavior which can be described well by the different types of rheological models. This study investigates the onset of gelation behavior of partially hydrolyzed polyacrylamide-hexamine-pyrocatechol polymer gel system which may be used to control excessive water production in the oil fields. Rheological measurements of this system have been performed at different time intervals and pH at 90°C. Attempts have been made to validate the onset of gelation behavior with Bingham plastic, Herschel-Bulkley, Mizrahi-Berk, and Robertson-Stiff model.

Results

It was observed that the developed polymer gel system under the present study has better agreement with the Robertson-Stiff model.

Conclusions

The viscosity of cross-linked polymer gel solution increases with temperature with the passage of time. This increased viscosity leads to gel formation which in turn behaves as flow diverting agent or blocking agent for controlling excessive water production in the oil fields.

BACKGROUND

The excessive water production in association with crude oil is one of the major production difficulties for the oil industries worldwide. The costs of lifting, handling, separation, and disposal of large amounts of produced water increase the operating cost of the crude oil production and decrease the economic life of a well. Therefore, there is a need to reduce excessive water production [1-4].

The polymer gel treatment is one of the most useful chemical methods to reduce water production [5-9]. Polymer gels are typically composed of a water soluble polymer and cross-linking agents which are dissolved in brine. After allowing sufficient time, the gelant solution sets into a semisolid mass and behaves as flow-diverting or blocking agent [10-12]. The selection of a polymer gel system for a given well treatment strongly depends on reservoir conditions such as temperature, salinity, hardness, and the pH of the water used for the preparation of the gelant [13-19]. Other parameters to be considered for the proper selection of a given polymer gel system include salinity of the formation water, permeability of the target zone, and the lithology of the formation [20, 21].

The different polymers used for the development of polymer gel in the oil fields are polyacrylamide with different degrees of hydrolysis (partially hydrolyzed polyacrylamide or PHPA) and polysaccharide such as xanthan biopolymer. These polymers can be cross-linked with metallic and organic cross-linkers to produce a three-dimensional polymer structure of the gel [22].

Gelation behavior is a fundamental parameter in oil-field application. The gelation behavior will determine the injection period and how deep into the formation the gel solution can be placed. For the study of gelation behavior, several methods like bottle testing method, sealed tube method, dynamic shear method (rheometer), and steady shear method (viscometer) are reported in the literatures [23]. Different equations have been used to describe the flow behavior of polymer gel system. The rheological behavior of polymer gel system is non-Newtonian, in which there is a nonlinear relationship between shear stress and shear rate. These polymer gel systems do not display simple behavior and require more complex models for their characterization [24].

In this work, the rheological behavior of PHPA, hexamine, and pyrocatechol gel solution was studied and was simulated using different rheological models. The rheological and gelation behaviors of polymer solution due to the effect of temperature and pH were evaluated using the equipment of Physica Rheolab MC1 (Anton-Paar GmbH, Stuttgart, Germany).

METHODS

Materials

The materials used in this work are partially hydrolyzed polyacrylamide, hexamine, pyrocatechol, sodium chloride, hydrochloric acid, and sodium hydroxide. Partially hydrolyzed polyacrylamide in the form of white crystalline powder (procured from Oil and Natural Gas Corporation Limited, Mumbai, India) was used as the water soluble polymer in carrying out the experimental work. The organic cross-linkers hexamine and pyrocatechol were obtained in solid form from renowned manufacturers: Otto Kemi Pvt. Ltd., Mumbai, India and Central Drug House (P) Ltd., New Delhi, India, respectively. Sodium chloride is used to maintain the salinity of polymer gelant, purchased from Nice Chemical Pvt. Ltd., Kerala, India. Hydrochloric acid (Central Drug House Ltd.) and sodium hydroxide (S. D. Fine–Chem Ltd., Mumbai, India) are used for the adjustment of pH.

Equipment

All rheological measurements were performed in a stress-controlled Physica Rheolab MC1. It is a rotational rheometer based on the Searle principle. It features a DC motor drive and optical encoder which provides excellent speed regulation, dynamic range, and transient response. These features provide superior performance in comparison with rheometers using gear-driven (indirect) or stepper motor drives. It enables precise measurements of fluid viscosity over the widest range of conditions available today. This machine is not equipped with a high pressure cell, and measurement at high temperature is not allowed. Hence, the gel had to be cured in hot air oven. The cured samples were transferred to the rheometer, and tests were conducted at 30°C. If these experiments were done at higher temperature, the viscosity of the polymer gel may have increased. Viscosity measurement tests were performed at different shear rates ranging from 1 to 1,200 s^{-1} using cone-and-plate geometry with a logarithmically increasing scale. Gelled samples were placed on the lower plate, and the upper spindle was brought slowly to the 0.2-mm gap. The reasons why a cone-and-plate sensor configuration was chosen as a test geometry are (1) cleaning is very easy after each measurement, (2) the cone and plate can be easily covered with tissue paper, and (3) there is a relatively smaller gap error due to a larger gap size between the cone and plate (0.2 mm in this experiment) compared to a cone-and-plate sensor (50-mm diameter, 1° angle).

Rheological Measurements

The steady shear flow properties of prepared partially hydrolyzed polyacrylamide-hexamine-pyrocatechol gel solutions were measured using a stress-controlled Physica Rheolab MC1. All measurements were performed at a fixed temperature of 30°Cover a wide range of shear rates from 1 to 1,200 s^{-1} with a logarithmically increasing scale. The 2 ml of the sample was used for each measurement of the polymer gel solution. The cone and plate were covered with tissue paper in order to remove a wall slippage between the test material and the cone and plate. Through a preliminary test using a direct visualization technique in which a straight line marker was drawn from the cone to the lower plate passing through the free surface of the sample solution, it was confirmed that a wall slip effect could almost be eliminated over a shear rate range tested by covering the plate surfaces with tissue paper. Special care was taken to minimize the effect of work softening when the sample solution was initially loaded on the plate each time. In all measurements, a fresh sample solution was used and rested for 20 min after loading to allow material relaxation and temperature equilibration. It was found from a preliminary test that 10 min of resting time is enough for all sample solutions to be completely relaxed and to be thermally equilibrated. All measurements were made at least three times for each test, and highly reproducible data were obtained within the coefficients of variation of ±5% in all cases.

RESULTS AND DISCUSSIONS

In the partially hydrolyzed polyacrylamide structure, the amide group ($CONH_2$) is converted to hydrophilic carboxylate group ($COONa^+$) by hydrolysis reaction, and further cross-linking agent builds up a complex network with carboxylate groups of polymer and forms a three-dimensional gel network structure.

The gel point is defined as the time needed to reach the inflection point on the viscosity vs. time curve (Figure 1). This method has been widely used by several authors [7, 16]. Before the gel formation, the viscosity of the gelling solution is relatively low; therefore, it can be measured accurately. After the gel formation, it will be hard to get an accurate viscosity value. Generally speaking, the strength of the cross-linking polymer solution is hard to measure and also hard to define. When the gel has reached a certain high strength, the gel will deform and is easily transferred in various mechanical viscometers, which will bring some damage to the gel performance. Unless this gelling solution allows gelation in the viscometer, the value of the measurement cannot represent its real strength.

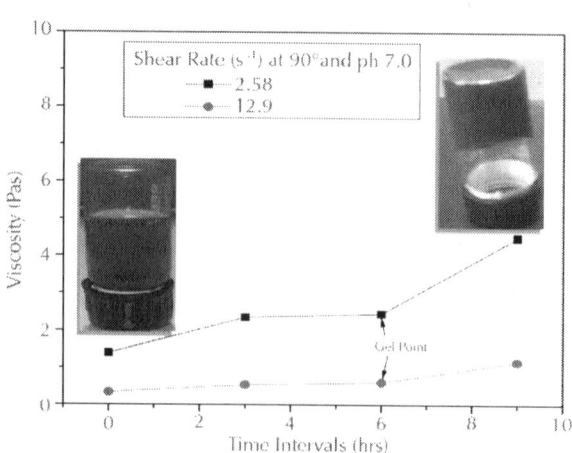

Figure 1: Change in viscosity with time intervals at constant shear rate, 7.0 pH, and 90°C. PHPA 0.9 wt. %, HMTA 0.4 wt.%, and pyrocatechol 0.4 wt.%.

The change in shear stress with shear rate at different reaction time intervals and pH was studied using Physica Rheolab MC1. It was observed that the polymer gel solution is non-Newtonian, and the values of shear stress increase with increasing shear rate which indicates that the viscosity of solution is shear rate-dependent. Also, it was observed that the values of shear stress increases with temperature at constant shear rate, which reflects that the viscosity of cross-linked polymer gel solution increases with temperature with the passage of time. This increased viscosity leads to gel formation which in turn behaves as flow-diverting agent or blocking agent for controlling excessive water production in the oil fields. There is no strong effect of pH on the rheological behavior of HMTA and pyrocatechol cross-linked partially hydrolyzed polyacrylamide gel system. Different polymer gel systems have different ranges of pH over which they can maintain their stability. All the experiments were carried out at pH 7.5.

Gelation Mechanism

The general mechanism involved in the polymer gel formation at any gelation temperature and pH (in this study, the temperature is 90°C and pH 7.5) is the cross-linking of polymer solution with organic or inorganic cross-linkers. In this case, the hexamethylenetetramine cross-linker is produced by the condensation reaction of formaldehyde with ammonia which then hydrolyses to form an unstable compound methanediol which dissociates to yield formaldehyde at room temperature. The hexamine cross-linker on hydrolysis yields formaldehyde which then combines with pyrocatechol cross-linker and forms 3, 4, 5, 6-tetramethylol pyrocatechol. Further, partially hydrolyzed polyacrylamide reacts with 3, 4, 5, 6-tetramethylol pyrocatechol and forms three-dimensional networks of polymer gel [25].

Applicability of Viscoplastic Flow Models

The developed polymer gel behaves like a non-Newtonian fluid; hence, viscoplastic flow models may be applied in this system.

Here, four inelastic-viscoplastic flow models including a yield stress parameter were employed to make a quantitative evaluation of the steady shear flow behavior of polymer gel solution, and then the applicability of these models was also examined in detail.

A general viscoplastic flow model having a yield stress term may be expressed by the following form:

$$\tau^{n_1} = \tau_o{}^{n_1} + k\gamma^{n_2}$$

(1)

where τ is the shear stress, τ_o is the yield stress, γ is the shear rate, k is the consistency index, and n_1 and n_2 are material parameters related to the material's flow behavior, and consequently, each flow model is determined according to the conditions of n_1 and n_2.

Table 1 summarizes the viscoplastic flow models adopted in this study and their flow characteristics. Here, τ_o is the yield stress of each model, and k is the consistency index related to a high-shear limiting viscosity, and γ_∞ as the shear rate is increased towards infinity. n is the flow behavior index, a material parameter that determines the shear-thinning nature of a material. In order that the models summarized in Table 1 could predict a shear-thinning behavior, both n_1 and n_2 must be positive values, and n_1 should be larger or equivalent than n_2.

Table 1: Viscoplastic flow models used in this study and their characteristics

Flow model	Equation	n_1	n_2	γ_∞	Shear-thinning condition
Bingham plastic	$\tau = \tau_o + k\gamma$	1	1	k	-
Herschel-Bulkley	$\tau = \tau_o + k\gamma^n$	1	n	0 k, if n = 1	$0 < n \leq 1$
Mizrahi-Berk	$\tau^{1/2} = \tau_o{}^{1/2} + k\gamma^n$	0.5	n	0 k^2, if n=0.5	$0 < n \leq 0.5$
Robertson-Stiff	$\tau = k(\gamma + C)^n$	1	n	k, if n = 1	$0 < n \leq 1$

Yadav and Mahto

Yadav and Mahto International Journal of Industrial Chemistry 2013 4:8 doi: 10.1186/2228-5547-4-8

In order to determine the polymer gel solution parameters of each model, a linear regression analysis was used for the Bingham model, while a nonlinear regression analysis method was used for the Herschel-Bulkley, Mizrahi-Berk, and Robertson-Stiff models. The calculated values of the polymer gel solution parameters from the viscoplastic flow models along with those of the determination coefficients and its comparison with the experimental values are reported in Tables 2, 3, 4, and 5 and Figures 2, 3, 4, and 5.

Table 2: Calculated flow model parameters for polymer gel solution (PHPA 0.9 wt.%, HMTA 0.4 wt.%, and pyrocatechol 0.4 wt.%) at 0 h, pH 7.5, and 30°C

Flow models	Parameters			
	τ_0 or C	$k(Pa \cdot s^{n-1})$	n	R^2
Bingham plastic	4.6845	0.0068	-	0.8671
Herschel-Bulkley	2.130	0.2704	0.5018	0.7942
Mizrahi-Berk	2.4372	0.0994	0.4136	0.8510
Robertson-Stiff	205.32	0.136	0.631	0.8788

Yadav and Mahto

Yadav and Mahto International Journal of Industrial Chemistry 2013 4:8 doi: 10.1186/2228-5547-4-8

Table 3: Calculated flow model parameters for polymer gel solution (PHPA 0.9 wt.%, HMTA 0.4 wt.%, and pyrocatechol 0.4 wt.%) after 3 h at 7.5 pH and 90°C

Flow models	Parameters			
	T_0 or C	$K (Pa \cdot s^{n-1})$	n	R^2
Bingham plastic	7.1737	0.0146	-	0.9563
Herschel-Bulkley	4.5052	0.1399	0.6937	0.9285
Mizrahi-Berk	4.8104	0.0561	0.5399	0.9512
Robertson-Stiff	218.571	0.111	0.747	0.9631

Yadav and Mahto

Yadav and Mahto International Journal of Industrial Chemistry 2013 4:8 doi: 10.1186/2228-5547-4-8

Table 4: Calculated flow model parameters for polymer gel solution (PHPA 0.9 wt.%, HMTA 0.4 wt.%, and pyrocatechol 0.4 wt.%) after 6 h at pH 7.5 and 90°C

Flow models	Parameters			
	T_o or C	K (Pa·s^{n-1})	n	R^2
Bingham plastic	7.2198	0.0181	-	0.9544
Herschel-Bulkley	2.5593	0.3632	0.5882	0.8785
Mizrahi-Berk	3.4865	0.1097	0.4828	0.9313
Robertson-Stiff	155.268	0.131	0.751	0.9554

Yadav and Mahto

Yadav and Mahto International Journal of Industrial Chemistry 2013 4:8 doi: 10.1186/2228-5547-4-8

Table 5: Calculated flow model parameters for polymer gel solution (PHPA 0.9 wt.%, HMTA 0.4 wt.%, and pyrocatechol 0.4 wt.%) after 9 h at pH 7.5 and 90°C

Flow models	Parameters			
	T_o or C	K (Pa·s^{n-1})	n	R^2
Bingham plastic	14.118	0.0224	-	0.9398
Herschel-Bulkley	7.0577	0.7382	0.5172	0.8368
Mizrahi-Berk	8.1910	0.1232	0.4640	0.8939
Robertson-Stiff	232.539	0.33	0.664	0.9489

Yadav and Mahto

Yadav and Mahto International Journal of Industrial Chemistry 2013 4:8 doi: 10.1186/2228-5547-4-8

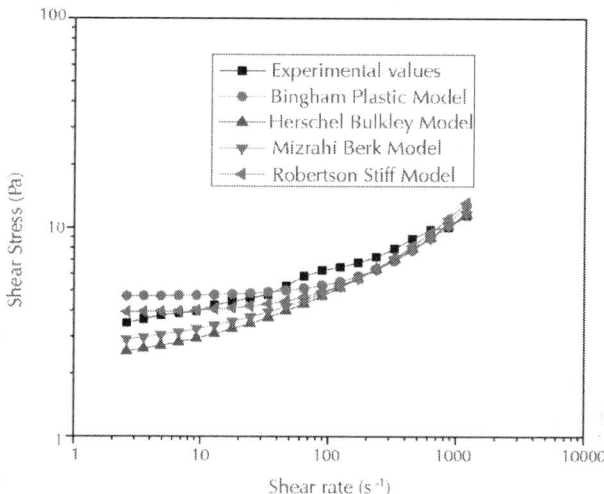

Figure 2: Rheogram of polymer gel solution (PHPA 0.9 wt.%, HMTA 0.4 wt.%, and pyrocatechol 0.4 wt.%).After 0 h at 7.5 pH and temperature 30°C.

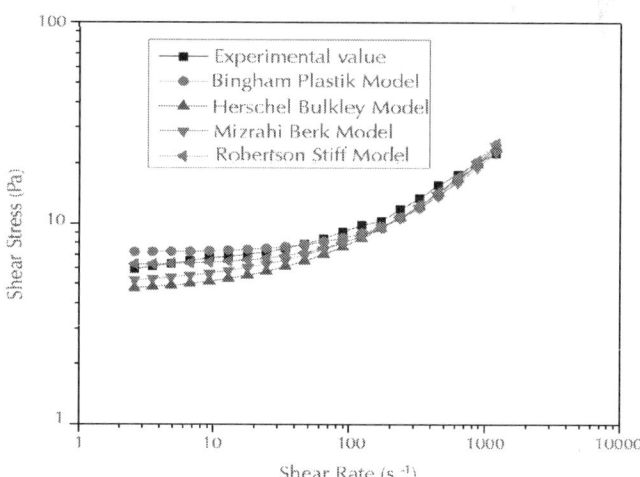

Figure 3: Rheogram of polymer gel solution (PHPA 0.9 wt.%, HMTA 0.4 wt.%, and pyrocatechol 0.4 wt.%).After 3 h at 7.5 pH and temperature 90°C.

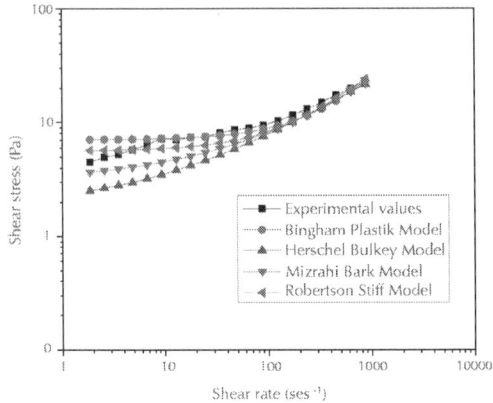

Figure 4: Rheogram of polymer gel solution (PHPA 9,000 ppm, HMTA 4,000 ppm, and pyrocatechol 4,000 ppm).After 6 h at 7.5 pH and temperature 90°C.

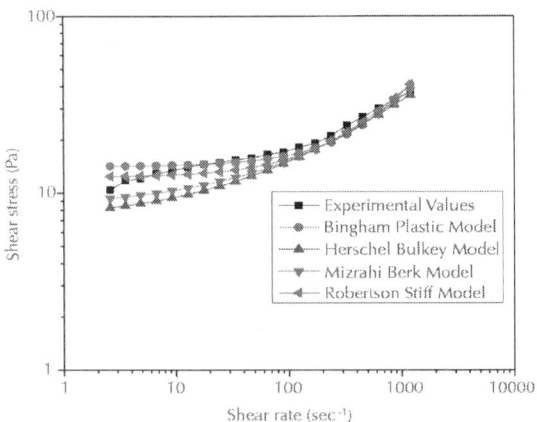

Figure 5: Rheogram of polymer gel solution (PHPA 9,000 ppm, HMTA 4,000 ppm, and pyrocatechol 4,000 ppm).After 9 h at 7.5 pH and temperature 90°C.

All the polymer gel solutions exhibit a marked shear-thinning behavior regardless of their pH. As expected, both the yield stress and consistency index values are not the same for the flow models. The value of the flow behavior indices provides a reference for the assessment of a shear-thinning nature, and it was observed that developed gelant is shear-thinning in nature.

The Bingham plastic, Herschel-Bulkley, Mizrahi-Berk, and Robertson-Stiff models give an excellent ability to describe the flow behavior of polymer gel solution. All of these four models have the values to determine coefficients near to 1 and show a clear tendency that the yield stress values remain unchanged with change of pH of the polymer gel solution.

The similar values of the determination coefficients are obtained for different flow models. The Bingham plastic, Herschel-Bulkley, Mizrahi-Berk, and Robertson Stiff models are all in good agreement with the experimentally measured data over a whole range of shear rates tested. But Robertson stiff model has better agreement with the experimental values of shear stress and shear rate data of the developed polymer gel system than the other models tested.

EXPERIMENTAL

Initially, the stock solution of partially hydrolyzed polyacrylamide in brine solution (1.0 wt.%) was prepared and kept for aging at ambient temperature for 24 h. Further, fresh solutions of hexamine and pyrocatechol were prepared in brine ahead of mixing the polymer and cross-linkers to form gelant. The appropriate solution of partially hydrolyzed polyacrylamide, hexamine, pyrocatechol, and brine were thoroughly mixed at room temperature by magnetic stirrer. The pH of the gelant solution was measured by Century CP-901 digital pH meter (Century Instruments Private Limited, Chandigarh, India). The pH of the gelant was maintained using 1N sodium hydroxide and 1N hydrochloric acid. Finally, the solution was transferred into small glass bottles and kept at the desired temperature in the hot air oven (90°C). At regular intervals of time, rheological behavior at different shear rates was measured using Physica Rheolab MC1. The concentration of polymer was 0.9 wt.%, whereas those of the cross-linkers, i.e., pyrocatechol and HMTA, were 0.4 and 0.4 wt.%, respectively.

CONCLUSIONS

The following conclusions are drawn from the present study:

- The developed polymer gel solution behaves like a non-Newtonian shear-thinning fluid.
- The rheological behavior of polymer gel varies with time as the gelation time increases; the gel solution will be more viscous, and the increase in viscosity is due to the cross-linking reaction.
- The nonlinear rheological models like Herschel-Bulkley, Mizrahi-Berk, and Robertson-Stiff models are found to be more suitable compared to the linear rheological model like the Bingham plastic flow model.
- Out of the three nonlinear rheological models (Herschel-Bulkley, Mizrahi-Berk, and Robertson-Stiff), the Robertson-Stiff model has better agreement with the experimental values of shear stress and shear rate data of the developed polymer gel system.

AUTHOR'S CONTRIBUTIONS

USY carried out the experiments using the Physica Rheolab MC1 and analyzed the data with non-Newtonian viscoplastic flow models using nonlinear regression method. VM has supervised the research work. USY and VM participated in the interpretation of results and drafted the manuscript. All authors read and approved the final manuscript.

ACKNOWLEDGMENTS

The authors are very thankful to the Council of Scientific and Industrial Research (CSIR), New Delhi, India for the financial support to carry out the research work.

REFERENCES

1. Urbissinova TS, Trivedi J, Kuru E (2010) Effect of elasticity during viscoelastic polymer flooding: a possible mechanism of increasing the sweep efficiency. J Can Pet Technol 49:49–56

2. Stokke BT, Elgsaeter A, Smidsrod O, Christensen BE (1995) Carboxylation of scleroglucan for controlled crosslinking by heavy metal ions. Carbohydr Polym 27:5–11

3. Moradi-Araghi A (2000) A review of thermally stable gels for fluid diversion in petroleum production. J Pet Sci Eng 26:1–10

4. Yadav US, Mahto V (2013) Modeling of partially hydrolyzed polyacrylamidehexamine-hydroquinone gel system used for profile modification jobs in the oil field. Journal of Petroleum Engineering. doi:10.1155/2013/709248

5. Mortensen K, Brown W, Jorgensen E (1994) Phase behavior of poly (propylene oxide)-poly (ethylene oxide)-poly (propylene oxide) triblock copolymer melt and aqueous solutions. Macromolecules 27:5654–5666

6. Perez D, Fragachan FE, Barrera RR, Feraud JP (2001) Applications of polymer gel for establishing zonal isolations and water shutoff in carbonate formations. SPE Drill Completion 16:182–189

7. Chatterji J, Borchardt JK (1981) Applications of water-soluble polymers in the oil field. J Petrol Technol 33:2042–2056

8. Al-Muntasheri GA, Nasr-El-Din HA, Hussein IA (2007) A rheological investigation of a high temperature organic gel used for water shut-off treatments. J Pet Sci Eng 59:78–83

9. Woodcock JW, Jiang X, Wright RAE, Zhao B (2011) Enzyme-induced formation of thermo reversible micellar gels from aqueous solutions of multiresponsive hydrophilic ABA triblock copolymers. Macromolecules 44:5764–5775

10. Kavanagh GM, Ross-Murphy SB (1998) Rheological characterization of polymer gel. Prog Polym Sci 23:533–562

11. Taylor KC, Nasr-El-Din HA (1998) Water-soluble hydrophobically associating polymers for improved oil recovery: a literature review. J Pet Sci Eng 19:265–280

12. Needham RB, Doe PH (1987) Polymer flooding review. J Petrol Technol 39:1503–1507

13. Storm ET, Paul JM, Phelps CH, Sampath K (1991) A new biopolymer for high-temperature profile control: part 1 - laboratory testing. SPE Reservoir Eng 6:360–364

14. Wever DAZ, Picchioni F, Broekhuis AA (2011) Polymers for enhanced oil recovery: a paradigm for structure–property relationship in aqueous solution. Prog Polym Sci 36:1558–1628

15. Cai W, Huang R (2001) Study on gelation of partially hydrolyzed polyacrylamide with titanium (IV). Eur Polym J 37:1553–1559

16. Grattoni CA, Al-Sharji HH, Yang C, Muggeridge AH, Zimmerman RWJ (2001) Rheology and permeability of crosslinked polyacrylamide gel. J Colloid Interface Sci 240:601–607

17. Al-Muntasheri GA, Hussein IA, Nasr-El-Din HA, Amin MB (2007) Viscoelastic properties of a high temperature cross-linked water shut-off polymeric gel. J Pet Sci Eng 55:56–66

18. Seright RS, Martin FD (1993) Impact of gelation pH, rock permeability, and lithology on the performance of a monomer-based gel. SPE Reservoir Eng 8:43–50

19. Mortimer S, Ryan AJ, Stanford JL (2001) Rheological behavior and gel-point determination for a model Lewis acid-initiated chain growth epoxy resin. Macromolecules 34:2973–2980

20. Park EK, Song KW (2010) Rheological evaluation of petroleum jelly as a base material in ointment and cream formulations: steady shear flow behavior. Arch Pharmacal Res 33:141–150

21. Scott T, Roberts LJ, Sharpe SR, Clifford PJ, Sorbie KS (1987) In-situ gel calculations in complex reservoir systems using a new chemical flood simulator. SPE Reservoir Eng 2:534–646

22. Hao J, Weiss RA (2011) Viscoelastic and mechanical behavior of hydrophobically modified hydrogels. Macromolecules 44:9390–9398

23. Ohen HA, Bick EF (2011) Golden section search method for determining parameters in Robertson-Stiff non-Newtonian fluid model. J Pet Sci Eng 4:309–316

24. Kelessidis VC, Maglione R (2006) Modeling rheological behavior of bentonite suspensions as Casson and Robertson–Stiff fluids using Newtonian and true shear rates in Couette viscometry. Powder Technol 168:134–147

25. Yadav US, Mahto V (2012) Experimental studies, modeling and numerical simulation of gelation behavior of a partially hydrolyzed polyacrylamidehexamine-pyrocatechol polymer gel system for profile modification jobs. International Journal of Advanced Petroleum Engineering and Technology 1:1–16

Advancement of Hydro-Desulfurization Catalyst and Discussion of Its Application in Coal Tar

Zongkuan Liu, Lei Zhang, Jian Jiang, Cheng Bian, Zichao Zhang, and Zifeng Gao

Chemical Engineering Department, School of Chemical Engineering, Xi'an Jiaotong University, Xi'an, China

ABSTRACT

This paper describes the influences of active metal, promoter and chelating agent on the properties of hydro-desulfurization catalyst. The use of chelating agent, especially its combination with common promoters e.g., EDTA-P, has an important meaning to develop highly active catalyst, specifically to unify the active metal dispersion degree and sulfurization degree in some extent, however, they are contradictory in conventional cognition. In the aspect of carriers, composition and nanometer carriers have more excellent performances in acidity, pores structure and metal-carrier interaction than common carriers, and are the developing trend in the future and should be a breakthrough mainly in preparation methods. We also pointed out the decisive factors to improve the activity of the catalyst: higher sulfurization degree of active metal oxide and higher aspect ratios of active phase crystal morphology, and the proper acidity and pores structure can be considered the key factors for deep desulfurization whose mainly obstacle is the desulfurization of large rigid molecules, e.g., dibenzothiophene and 4, 6-dimethyl substituted dibenzothiophene. Based on above that, We discussed the suitable hydrodesulfurization (HDS) catalyst for coal tar, aiming at providing some theoretical guidance for the "design" of coal tar HDS catalyst.

INTRODUCTION

China's current energy situation is rich in coal, lack of natural gas and less of oil. In recent years, with the development of the national economy, the contradiction between supply and demand of oil and gas resources is more and more prominent. Using coal tar to prepare fuel oil will alleviate the energy crisis in a certain extent. Coal tar is the main byproduct of coal gasification and coal coking process, its output was about 15 million tons with the annual growth rate about 13.68% in 2009 in China [1, 2].

Hydrotreating is the main technology to prepare fuel oil from coal tar, and hydrodesulfurization (HDS) is an important link in the process of which core is HDS catalyst. Traditional HDS catalyst does increasingly not meet the requirements and the following two reasons could account for it: on one hand, coal tar is heavy and the unsaturated compounds which are abundant in coal tar have evil influences (resinification reaction reacts between unsaturated compounds and chemical reagent, and thus creating much precipitation and slagging) on coal tar hydro-refining [2]; on the other hand, the environmental protection laws are more and more strict with sulfur content of fuel oil. Pawelec et al. [3] pointed out that the activity of the new catalyst should be 3.2 times the conventional catalyst when requiring diesel sulfur content reduces from 500 ppm to 15 ppm with not changing the technological parameters. It is press for designing and developing new HDS catalyst. HDS catalyst includes loaded type and un-loaded type, this paper mainly studied on the former.

The catalyst design process is very complex. In the past, preparation of catalyst depends on the rich "formula" experiences; in recent years, along with the development of surface physics, surface chemistry and organic catalytic mechanism, as well as the use of precision instrument in catalyst characterization, making catalyst "design" be possible. The components that can be distinguished in the catalyst include active metal, promoter or chelating agent and carrier.

Active metal is the main source of hydrogenation activity, it should be chosen according to the properties of raw oils or model molecules and product quality standards [4-12]. Promoter is an important part of the catalyst and a lot of confidential materials are concentrated in it. It is essential to optimize its kinds and content in the process of catalyst preparation [13-22]. The available carriers in HDS catalyst are as the following types according to the chemical compositions: silicon-aluminum molecular sieves [23,24], metallic oxides [25,26], mesoporous silica molecular sieves [10,11,27,28] and composites made up by these materials [7,8,23,26]. The most significant characteristics of carriers are its ability to disperse metal and possession of B, L type acids. These two kinds of acids play different roles in hydrogenation reaction, but synergy exists between them [29].

The three parts of catalyst are interrelation and interaction. Hence, a comprehensive consideration should be given in the process of catalyst preparation. In the actual process of catalyst "design", "Attending to one thing and losing sight of another" is a frequently encountered problem. For one instance, weak metal-carrier interaction is beneficial to the improvement of sulfurization degree of the active metal oxide, but the dispersion degree of the active phase may be reduced. For another instance, although large pores can weaken the diffusion resistance and reduce secondary cracking of the reactant molecules, that would make the initial activity of the catalyst and the metal capacity of carrier reduced. Consequently, the problem of the catalyst "design" is actually an "optimization" problem. The objective function should be the maximization of enterprise's interests, the constraint conditions should be the nature of raw oils, quality standards of products and laws of environmental protection, and the optimization variables are main parameters of catalyst design: metal types, metal content and crystal morphology of active phase, promoter or chelating agent types and content, and carrier's surface properties. The main purpose of this paper is to provide certain theoretical guidance for the preparation of coal tar HDS catalyst on the basis of system analysis and introduction of the main parameters of the HDS catalyst.

ACTIVE COMPONENTS

Conventional Active Components

In general, conventional hydrogenation active sites include the simple substance of noble metal and non-noble metal sulfide. But the use of noble metal is limited to some extent because of its expensive price. So, keep the activity of noble metals catalyst not decline while making the consumption of noble metals reduced is the goal that researchers always pursue. In this respect, Nakamura, et al. [30] proposed a nanometer-catalyst technology to reduce the dosage of noble metals. Its specific process was schematically shown in Figure 1. Its basic concept is that the catalyst is made up of the substrate to restrain the sintering of noble metal and the separation material to inhibit the agglutination of the substrate. The structure can greatly reduce the dosage of noble metal because it can keep noble metal be nanometer particle. To make an appropriate bonding force between metal and carrier is its key point, because it is not conducive to improving the catalytic activity whether the bonding force is too strong or too weak, and thus selecting a proper carrier is important. Weak resistance capability to sulfur is the other reason resulting in its limited application. Some studies [8,9] showed that the resistance capability to sulfur can be enhanced when catalyst possesses smaller metal particles, proper acidity (mainly Lewis acid sites with high electronegativity) and formation of alloy. This may relate to the inhibition of generation of low activity Pd_4S in these circumstances.

Co-Mo-S, Ni-Mo-S and Mo-Ni-W-S are commonly used active phases of non-noble metal HDS catalysts. In general, their hydro-refining activity increases as the sequence [4,5,31,32]. This may because different promoter metals (Co, Ni) have different modulation ability to S-metal bonding energy. Literature [3] confirmed this view and pointed out that the change process of the activity of Co(Ni)Mo(W)S is a volcano curve with the enhancement of S-metal bonding energy. Therefore, moderate S-metal bonding strength is one of the crucial factors to improve the catalytic activity.

Non-Noble Metal Phosphide and Noble Metal Phosphide

Metal phosphides are commonly obtained through the reduction of their corresponding oxides or chlorides by temperature-programmed method in H_2 atmosphere [33, 34]. The reduction temperature affects the particle size of active phase and hence has significant implications to HDS activity; therefore, selecting an appropriate reduction atmosphere to lower the reduction temperature is momentous.

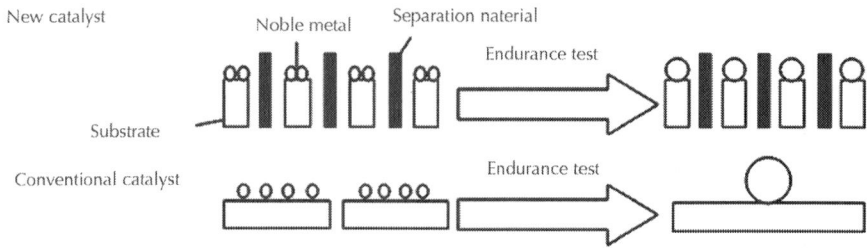

Figure 1: The new catalyst design concept about restraining sintering of the noble metal.

Non-noble metal phosphides have highly HDS activity which is higher than normal Co(Ni)Mo(W)S. Particularly, the activity of crystalline Ni_2P is the highest among them because of its higher intrinsic activity and dispersion degree compared to the other phosphides [10]; Another advantage of Ni_2P is its HDS activity will not be inhibited with the competition of HDN (hydrodenitrification) [4,6,10]. Inversely, the HDS activity of Co(Ni)Mo(W)S would be restricted due to the competition of nitrogen compounds in H and the edges of active phase [35]. The disadvantage of Ni_2P is its very low dispersion degree, only 1/3 times the active phases of industrial hydrotreating catalyst. Consequently, to improve its dispersion degree is the key to its practical application. JI research group [13-16] have studied the influences of W, Mo, Li, Na, K, Mg, Ca, Sr, Ba and B on the dispersion degree of Ni_2P based on SBA-15, manifesting that W, Mo, Ca and B promoted its dispersion degree, and thus the HDS activity of the catalyst was enhanced, but that property is closely related to the promoters content (as the discussion in 5.1.3 in this paper); In addition, Ni/P molar ratio is also a major parameter in process of the catalyst preparation because of its affection on the types of active phases. If the ratio is too high, there will be more formation of low active $Ni_{12}P_5$. In general, this ratio is between 1.2 - 1.4 [6,13,14,36], and is lower than the ideal ratio 2.

Recently, it is worth mentioning that Kanda et al. [33] synthesized a series of noble metal phosphides: Rh-P, Pd-P, Ru-P, Pt-P. They made the corresponding noble metal chloride precursors primarily decomposed in N_2 atmosphere, and then reduced by temperature-programmed method in H_2 atmosphere. Their HDS activity for thiophene follows the order: Rh-P > Pd-P > Ru-P > Pt-P, activity and stability of Rh-P are the best and are better than Ni-P.

Non-Noble Metal Carbides and Nitrides

Chemical bond in carbides and nitrides has part of the properties of metal bond. Their preparation methods mainly differ in the reaction atmosphere and temperature [12]. Nitrides are commonly generated in NH_3 and N_2 atmosphere, the former's advantage over the latter's is to inhibit the formation of impurities and lower the reduction temperature. The formation process of carbides is the oxide precursors reduced primarily by temperatureprogrammed method in CH_4-H_2 atmosphere, and then utilize N_2 containing low quantity of O_2 of which volume fraction is less than 1% to passivate the carbides. One of the remarkable properties of metal carbides is to embellish the acidity of silica-aluminum molecular sieves [20, 21].

Others

Lan et al. [4] made Ni_2P and Mo-Ni-W combined catalyst Ni_2P-NiMoW/ -Al_2O_3 which showed more excellent performances than conventional catalysts, that is, it can make diesel oil achieve deep desulfurization, simultaneously, the increase of cetane number.

ReS_2 also can be used for hydrogenation active site of HDS catalyst. Its HDS activity is positive correlation to the carrier's ratio of silica to aluminium, i.e., the acidity of the catalyst plays an inhibitory role for HDS activity [7], but oppositely for traditional HDS catalysts [28,37].

THE CONTENT OF ACTIVE COMPONENTS

Metal content including total metals content and the proportion of different metals in one catalyst is an important parameter in catalyst design process. Metal content usually affects the dispersive state of active metal, and the dispersive state would also impact the acids distribution [38], metal-support interaction [39] and the number of hydrogenation active sites etc. Overall, a highly active catalyst should be higher total metals content and appropriate metal dispersion degree, i.e., in the case of the excellent dispersion state of active metals, the higher total metals content is conducive to improving the catalytic activity.

Influences on Acids

In normal circumstances, the appropriate acids distribution and moderate acidity strength are beneficial to HDS activity of catalyst. That is because, on one hand, B, L acids are in favor of the adsorption of reactant molecules [28] and hydrogen overflow, and thus are useful to cleave C-S [37,40]; on the other hand, the acids can make the alkyl-substituted dibenzothiophene (DBTs) isomerize to molecules owning weaker steric hindrance for -S-.

There are three regulation mechanisms of metal active on acid amount and acids type: to generate new acid, ion exchange and covering effect. Acidity is changed as the change of metal content. When the active metal content increases within a certain range, the amount of L acid (weak acid) is increased but inversely for B acid [4,41]. The increase of L acid amount is due to the empty orbits in active metals which can generate Lewis acid that is different from the Lewis acid site centering in Al_3^+; the reduction amount of B acid is because of ion exchange between metal cation and proton, along with covering effect of metal components on part of Si-OH-Al (strong B acid center) and OH-Si (weak B acid site); Study

[28] on the acids distribution of NiW/Al-SBA-15 and AlSBA-15 proved this point, the former's L, B acid amount are 1.6, 0.5 times the latter, respectively. Study in literature [38] also supported this view. But literature [42] showed that the active metal also made B acid amount increase, this may relate to the conversion of tetra-coordinate metal species on the support surface [43].

The Consideration for Metals Content Ratio

There is synergistic effect in multi-components active metal catalyst, and the effect can make catalyst be better performance. e.g., HDS activity of PdAu/HS-HMS for thiophene was three times Pd/HS-HMS due to the synergistic effect between Pd and Au [9], and the deactivation rate of the former was also significantly lower than the latter. The main reason is that the former has the formation of the alloy made up of Pd and Au, and thus inhibits formation of PdS_4. The synergy closely relates to the metals content ratio. Consequently, both the total content and ratios of the metals content should be taken into account in the process of catalyst design. Literatures [44, 45] discussed the impact of molar ratios of Pt/(Pt + Pd) and Ni/(Ni + W) on the catalytic activity, demonstrating that the former or the latter possessed the highest hydrogenation activity for naphthalene or HDS activity for thiophene when the ratios are 0.7 or 0.28, respectively. Study on Mo-Ni$_2$P/SBA-15/ cordierite monolithic catalyst [16] indicated that its HDS activity for DBT was optimal when the molar ratio of Mo/Ni + Mo was 0.18, however, when the ratio was 0.26 in NiMo/Al$_2$O$_3$-nmY, the optimal desulfurization degree of diesel was up to 99.6% [23]. Through the latter two examples, we can find that the ratios of Mo/Ni + Mo are different, that may because the active phases are different in the two catalysts. So, the ratio is not invariable, and should be optimized in accordance with the active phase, carrier and preparation method.

THE ACTIVE PHASE CRYSTAL MORPHOLOGY

The active phase crystal morphology has vital influence on the activity of the catalyst. It is closely relevant to the types and amount of the active phases, as well as the number of the corner and edge active sites and lattice defects. In addition, it can also determine whether the hydrogenation active sites can be well exposed to the reactant molecules or not. The main parameters used to describe the active phase crystal morphology include stacking degree, length and curvature of the lamellae.

Stacking degree has major impact on the types of active phase and the amount of corner and edge active sites. Generally speaking, multilayer stacking shows better performance than the single-layer. This is mainly manifested in two aspects: First, multilayer stacking can generate more type II active sites than the single-layer due to its weaker interaction with carrier except the basal layer [46]. The top layer possesses the highest activity because of the weakest steric hindrance and higher unsaturation degree of the brim and corners active sites [8], in addition, the top layer and middle layers also differ in the desulfurization mechanism, the brim of the top layer owns capability of DDS (direct desulfurization) and HY D (hydrogenation desulfurization) routes while the middle layers only have DDS activity [47]. Second, multilayer stacking can produce more lattice defects compared to monolayer stacking and thus is beneficial to chemisorption [26,28]. However, the number of stacking layers is not the more the better; too many stacking layers will lead to a ratio decrease of vertical adsorption, as well as the corners and edges active sites [8,39,47]. Therefore, it is major to optimize stacking layers in order to balance the amount of corners and edges active sites and II type active sites.

However, the multi-layer stacking is not the necessary condition to produce type II active sites. Sometimes, monolayer stacking may be the II active sites. Parola et al. [48] pointed out the prerequisite of generating type II active sites is weaker metal-carrier interaction.

Moreover, in case of inherently weak metal-support interaction, the increase of active phase stacking layers, sometimes, leads to a detriment effect on its HDS activity [28], Study [8] about NiMo-NTA/Al$_2$O$_3$(NTA = nitrilo triacetic acid that can significantly weaken the metal-carrier interaction) showed that its intrinsic HDS activity for DBT significantly reduced with the increase of active phase stacking layers because of activity of the top layer higher than others, and more stacking layers may mean lower proportion of the top layer [26]. Nevertheless, when the stacking layers are few, it is not conducive to the plane adsorption of reactant of catalyst [25,28]. Therefore, even in the case of weak intrinsic metal-support interaction, and hence generating a large number of type II active sites, an optimal point of the stacking layers should also exists, rather than the fewer the better.

In short, stacking layers of the active phase has important influences on HDS activity. However, the relation between active phase crystal morphology and HDS activity can't be determined only depending on the stacking degree of the crystal morphology. Shimada et al. [46] summarized the relation, indicating that the higher aspect ratios (approximately equal to the thickness divided by length) of the crystal morphology, the higher intrinsic activity of the catalyst. Many studies [5,23,26,28,32,49] have supported the view. That may because this kind of crystal morphology can maximize the amount of corner and edge active sites, and weaken steric hindrance of the active sites so that it could be well exposed to the reactant molecules, as well as generate more type II active sites. Finally, we can also safely get such a conclusion according to those literatures: moderate metal-support interaction is the decisive factor for achieving of the highest aspect ratios of crystal morphology.

The bending of the active phase lamellae also has an important impact on HDS activity and HDS mechanism [47]. This is because, on one hand, the curved lamellae would make the edge of metal layer expose more active sites, on the other hand, it make the S-Mo (W) bond tight ,and thereby increasing the unsaturation degree of the active metal and hence creating new active sites.

PROMOTERS AND CHELATING AGENTS

Promoters

The modulation capability of promoters for catalyst performances is not only relevant to the nature of the promoter itself and quantity of the promoter, but also connected with the impregnation order of active metal and the promoter [18, 19, 22, 50, and 51]

Influences on Metal-Support Interaction

Metal-support interaction is the decisive factor to the sulfurization degree and crystal morphology of active metal. Therefore, utilizing promoter to modulate metalsupport interaction is important. In general, the modulation of the metal-support interaction actually refers to "weaken" this interaction.

Metal-support interaction is modulated by promoter mainly in three ways: First, it can be weakened by modulating the coordination state of active metal. Different coordination states of the active metal bring about different intensity of metal-support interaction, more specifically, tetrahedron coordinated active metal species commonly have stronger binding force with support than octahedron species [26,28]. P modulates the metal-carrier interaction is just through that mechanism [19]. But the improvement degree of metal-support interaction is mainly associated with the impregnation sequence of P [18], to be specific, co-impregnation improvement capacity is higher than the surface impregnation; second, metalsupport interaction is modulated through synergy effect between the active metal and promoter, the promoters mainly should be transition metals, e.g., Fe [50], Co [51]. To these promoters, The content must be suitable because excessive content of promoter may inversely strengthen the metal-support interaction due to the improvement of the dispersion degree of active phase in this case [50]; third, promoters could achieve the modulation ability through "isolation" of the active metal and support in physical space. For B modified catalyst B-MoNi/ Al_2O_3 and MoNi-B/Al_2O_3 [22], the former which impregnated B and MoNi in sequence has better modulation capability than the latter. This is because B (w0.6%) could be monolayer distribution on the carrier surface in B-MoNi/Al_2O_3, which "separates" active metal and carrier in some extent and thereby weaken the metal-carrier interaction.

Influences on the Acids Distribution and Acid Amount

The major impacts of promoter on the acid amount and acids distribution are fulfilled through the following five kinds of mechanisms: first, it is modulated by ion exchange, mainly some metal promoters; second, the modulation is achieved in the way of promoting the dispersion degree of active metal; third, "covering" effect; fourth, the acidity also can be modulated by some promoters that are inherently acidic, such as transition metal ions; fifth, the modulation can be obtained by substitution reaction between skeleton aluminum and promoter. Sometimes, one promoter may have, simultaneously, several roles to modulate, therefore, to optimize its content and impregnation orders to get desired acidity is important.

Mg, K, P and W all own the ability to modulate the acidity of catalyst. Mg can promote the dispersion degree of metal component and hence make the metal surface present more electron-deficient sites; as a result, the number of L acid is increased. e.g., the density of L acid of (w) 2% Mg-(w) 7% Mo_2C/HY is nearly doubled compared to (w) 7% Mo_2C/HY [20]. K usually makes the amount of B acid significantly reduced due to proton exchange between hydroxyl hydrogen on carrier surface and K [20,21]. However, study [17] on CoMo/ -Al_2O_3 showed K mainly made L acid amount reduced. Study [18] about P-NiW/Y-SiO_2-TiO_2 showed that both surface impregnation and co-impregnation methods resulted in significantly reduction of the total amount of acid, but in the case of co-impregnation, although P covered part of acid sites on the carrier's surface, simultaneously, the dispersion degree of active metals was improved, so that the amount of L acid was increased. The reason for the increase of L acid amount is the same to Mg. In addition, P may substitute for the carrier's framework aluminum, more specifically, a framework aluminum hydroxyl of the zeolite would be substituted by two phosphorus hydroxyl, making B, L acid amounts modulated [24]. Transition metal ions such as W have empty orbits; their addition can make a well increase of the amount of acid [15, 45]

Influences on the Pores Distribution and Specific Surface Area

The realization of impact of promoters on pores distribution and specific surface area is mainly based on its influences on metal dispersion state.

Generally, with the increase of promoter content, the specific surface area and pore volume increase primarily and then decrease, i.e., there is an optimum point. For B-MoNi/Al_2O_3 [22], W-Ni_2P/ SBA-15 [15], Mo-Ni_2P/SBA- 15/cordierite [16], when B (w) of 0.8%, W (w) 3%, Mo (w) 4.2%, their specific surface area and pore volume are maximal, respectively. Such effect of B, W and Mo can be attributed to the impact of promoter content on the dispersion state of active metal. When the promoter content is low, the dispersion degree of the active metal is improved and thus the surface properties are ameliorated; on the contrary, if the content is too high, the active phase (or precursor) may aggregate together and block portion of the pores, resulting in the decrease of surface area and pore volume. But there are exceptions, e.g., Mg-Mo_2C/ HY [20,21], the pore volume and specific surface area primarily decrease and then increase with the increasing content of Mg.

Chelating Agents

Chelating agents are typically organic compounds with donor atom having two or more available electrons. They are capable of binding metal ions to form chelates, such as NTA [8,52], EDTA [53], and CyDTA etc. Their common feature is that they can delay the sulfurization of promoter metal (Co) until the active metals (Mo, W) are fully sulfurized, avoiding the production of Co_9S_8 phase that is thermodynamically unstable, which is conducive to the formation of the active phase of Co-W(Mo)-S. Another remarkably notable feature of the chelating agents is that they can dramatically weaken the metalsupport interaction, but scarcely any impact on the dispersion degree of active phase, or even improve the dispersion degree of active phase [49,52]. The nature of chelating agents breaks through the traditional cognition which the improvement of metal dispersion can be achieved only by enhancing the metal-support interaction, and makes sulfurization and dispersion degree of the active phase unified in a certain extent, providing more ways for people to prepare highly active catalyst. For instance, Yu et al. [49] prepared NiW/Al_2O_3, NiW-P/ Al_2O_3, $NiWP/EDTA/Al_2O_3$ three catalysts, their HDN activity for quinoline were improved as the order. The addition of EDTA in $NiWP/EDTA/Al_2O_3$ made the average number of the stacking layers and the length of active phase crystal morphology reduced 0.3 layers and 1.2 nm, respectively, with respect to $NiWP/Al_2O_3$, making the aspect ratios of NiWP/EDTA/ Al_2O_3 improved; moreover, the sulfurization degree of active metal is also pretty increased because of weak metal-carrier interacttion due to the addition of EDTA. If NiW/Al_2O_3 is modified directly by EDTA, it may cause too high dispersion degree of the active phase, which is not conducive to the molecules adsorption and is harmful to the increase of corner and edge active sites. Thus, P actually played a role like bridge in this experiment.

CARRIERS

In the preparation process of carrier, pore structure should be primarily design and topology analysis in order to balance the specific surface area and diffusion resistance (mainly determined by pores diameter) [11,40,54, 55]. Generally speaking, the relation between pore diameter and specific surface area is mutually restrictive. But, the principal restricted condition for HDS activity is different for different reactants. In terms of macromolecules, the relatively larger pore diameter is more important than larger specific surface area [55,56]. Secondly, choosing a suitable carrier having moderate binding force with active metal is crucial. When the binding force is too strong, It will lead to the formation of the Mo(W)-O-Al and thus the lower type I active site during sulfurization, whereas the formation of the type II active site requires Mo(W)-O-Al bond is broken (at least partially broken). Oppositely, when the interaction is too weak, there is no doubt that it is not conducive to improving the dispersion degree of the active site and hence does harm to the HDS activity [26].

The most commonly used carrier of HDS catalyst is -Al_2O_3 in industry. But it chiefly has two disadvantages, one is the dispersed pores distribution and short pores diameter concentrating in shorter than 5 nm resulting in large diffusion limit for reactant and intermediate products [11]; the other is that its strong interaction with active metal oxide. These shortcomings make it increasingly can't meet the requirements of desulfurization. The typically mesoporous silica molecular sieve materials SBA-15, HMS and MCM-41 as the carrier of hydrofining catalyst have potential value in use. Their advantages are large specific surface area (about 1000 m²/g), uniform controlled mesoporous (5 - 30 nm) and stable skeleton structure. Their disadvantages are weaker acidity which should be strengthen on the original basis in order to isomerize DBTs to molecules owning weaker steric hindrance for -S- [23,37], and weaker hydrothermal stability and mechanical strength, as well as too weak interaction with active metal. Those several factors limit its industrial application so that the modification study should be carried on which is commonly based on their surface Si-OH.

At present, the commonly used modification method of those disadvantages is to introduce Al [28,37,55], Ti [54], Zr [56], W [45], etc. into the carrier's frame. The methods to add heteroatom are usually co-impregnation (grafted directly) and post-grafting. the former's weaknesses are that the difficulty to introduce heteroatom in the case of low PH (lower than 2) [28], and to obtain larger and regular pores of catalyst [55,56]; its merit is it makes the catalyst possess higher specific surface area than post-grafting; The advantages of the latter are pore expanding and isomorphous substitution [55]; its shortcoming is pore plugging and thus making the specific surface area significantly lower than the catalyst obtained by co-impregnation method. The principle of post-grafting is firstly hydrolyzing the aluminum source (e.g. ammonium hexafluoroaluminate) to be $[Al(OH)_4]^-$, and then isomorphously substituting. Inspired by the princeple of post-grafting, Li et al. [28] developed a new method named high-temperature hydrothermal treatment (HTHT) to solve the pore blockage problem of post-grafting. $[Al (OH)_4]^-$ can be easily formed by Al^{3+} and Al-OH in condition of high-temperature hydrothermal; the other key of the method is that PH should be isoelectric point nearby to generate lots of Si-OH, and finally through the following reaction to complete graft.

Si-OH + HO-Al → Si-O-Al + H_2O Besides the preparation method, a crucial parameter need to be optimized is the content of heteroatom, which is closely related to the dispersion degree of active phase and acids distribution of catalyst etc. e.g., the HDS performance of NiMo/TiSBA-15 on coke light gas oil (KLGO) was the best when Si/Ti mole ratio was 20. Research of NiMo/Zr-SBA-15 (synthesized directly) also had similar result [56].

Along with more and more heavy oil and increasingly strict laws of environmental protection, single carrier based catalyst already cannot meet the demand. Composite carrier commonly shows more excellent HDS activity than single carrier, such as Al_2O_3-ZrO_2 [57], Al_2O_3-SiO_2 [40], TiO_2-SiO_2[31], Beta-SBA-15 [37], TiO_2-SBA-15 [25] and ZrO_2-SBA-15 [25,26,32,56]. In those carriers preparation process, the most important design parameter is the proportion of carrier contents in one composite carrier, because the proportion has an important relation with pores diameter, specific surface area, acids distribution and intensity, as well as metal-carrier interaction [31, 40]. e.g., ZrO_2-SBA-15, the optimal quality ratio of ZrO_2/SBA-15 is generally 1:3 [25,26,32]; as to SiO_2- Al_2O_3, when SiO_2/Al_2O_3

quality ratio was 1:10, the HDS activity and inactivation rate can reach a satisfactory balance [40].

It is greatly worth to mention that carrier on nanometer level has dramatically better performance on acidity, pore structure and carrier-metal interaction than carrier on micrometer level [23,58]. e.g., HDS degree of NiMo/ Al_2O_3-nmY was increased by 0.8% compared to NiMo/ Al_2O_3-umY, and the rate constant of the former is 3.46 times the latter [23]. In addition, the carriers on nanometer level can protect molecular sieve from poisoning and shorten diffusion path for raw material [58]. In total, its advantages are highly obvious. But, its preparation faces many problems, such as the need of expensive structure-directing-agent, and demanding strict reaction temperature in the synthesis process etc. In addition, the cordierite honeycomb ceramics monolithic catalyst [16] has the advantages of low bed pressure drop and high mass transfer efficiency, etc. over traditional hydrogenation catalysts, and hence also attracts researchers' attention From the above analysis, we can see that carrier's development trend is multicomponent and smaller particle size.

DISCUSSION ON DESULFURIZATION CATALYST THAT IS SUITABLE FOR LOW TEMPERATURE COAL TAR

Coal tars can be divided into three categories according to the coal pyrolysis temperature: high temperature coal tar, medium temperature coal tar and low temperature coal tar. The low temperature coal tar is the most suitable one to produce fuel oil because of its lower aromatics and asphalt content and higher alkane content. Typically, sulfur compounds in coal tar include thiol, sulfoether, disulfides, thiophene series and dibenzothiophene series. The commonly grading sequence of coal tar hydrotreating catalysts is: protection catalyst, demetallization catalyst, HDS catalyst, HDN catalyst, mild hydrocracking catalyst, which calls for hydrofining catalysts have some cracking ability. A pilot research [59] showed that catalysts used in heavy oil hydrotreating process are feasibly used for hydrotreating of low temperature coal tar. However, coal tar has its own characteristics; the following will theoretically discuss the suitable HDS catalyst for low temperature coal tar in the case of this kind of grading sequence.

First of all, its acid density and strength should be controlled, more specifically, the acidity should be stronger than demetallization catalyst and weaker than hydrodenitrification catalyst. That is because, on one hand, nitrogen content (0.48% - 1.13%) is high in coal tar [2], and the nitrogen with alkaline will poison the catalyst and inhibit HDS because of the competition effect of HDN; on the other hand, the moderate acidity can make HC and HDS well occur, moreover, can inhibit the formation of coke. Secondly, coal tar is a complex mixture, the sizes of sulfur compound molecules in coal tar are greatly different, that means these molecules need different degree of cracking and hydrogenation; moreover, aromatics content is very high in coal tar, and mainly tricyclic or above anthracene series. Those kinds of characteristics of coal tar require catalyst possesses different pores distribution and active sites distribution, namely, the activity in larger pores should be higher than in smaller pores in theory, which will match to the different reaction degree that is required by mixed molecular. That needs to explore more elaborate catalyst preparation methods. Thirdly, HDS catalyst is relatively near the front of grading sequence; hence, material flowing through HDS catalyst is very heavy, the catalyst should have powerful capacity of treating aromatics including aromatics adsorption and diffusion, as well as more effective hydrogenation and ring opening ability. This requires catalyst possesses more lattice defects, secondary pores and excellent synergy between hydrogenation and cracking sites, In addition, appropriate number of large pores (100 - 500 nm) is also important. This primarily need to develop a new metal loading concept to increase the quantity of lattice defects, and amount of secondary pores can be added through hydrothermal treating and synthesis of composite molecular sieve. Finally, coal tar contains much asphalt whose structure is very complex, the polycyclic aromatic hydrocarbons in asphalt are bonded by the sulfur bridge bond, aliphatic bond and metal porphyrin bond. The structures of nickel and vanadium porphyrin compounds are similar to polycyclic aromatic hydrocarbons in asphalt, leading to abundant poisonous metals exists in it, moreover, part of the asphalt would generate carbon deposit by condensation polymerization and graphitization reaction, these carbon deposit can jam the channel and cover the active sites of catalyst and thereby making HDS catalyst inactivation. To solve this problem, the hydrogenation and cracking sites should have well synergetic effect. For the problem of catalyst poisoning, it can be eased by adding some nanometer oxides in support [58].

CONCLUSIONS AND PROSPECT

- Transition phosphide, carbide and nitride catalyst possess potential practical value for oil hydrotreating. But there are also some drawbacks must be further improved. The main drawback of the phosphide is the lower dispersion degree; the use rate of unit mass catalyst related to the phosphide sites is low. Transition metal carbides and nitrides catalyst own higher initial activity for hydrorefining, but in the using process, the active phase will be gradually sulfurized, making the catalytic activity decline. To further overcome these shortcomings, it will make them become a new generation catalyst for hydrotreating. In addition, the combination of non-sul-fide metal catalyst with the sulfide catalysts may also be a research direction.

- Promoter or chelating agent can improve the active phase dispersion degree, acids distribution, pores distribution and metal-carrier interaction. Metal-carrier interaction is closely relevant to the crystal morphology of active phase. The using of promoter-chelating has important significance to develop novel catalyst preparation methods. We need to further develop new promoter (chelating agent), on one hand, to expand the catalyst preparation methods and to improve its thermal stability, mechanical properties and activity, on the other hand, to weaken the contradictions related to the performances of catalyst.

- Composite and nanometer carriers can provide more appropriate acidity, pore structure and metal-carrier interaction than conventional carriers, but its thermal stability and mechanical properties are need to be further improved. To develop new preparation method is the key for the use of the composite and nanometer carriers.

The catalyst "design" problem is an "optimization" problem, and a system analysis should be given in process of catalyst preparation. The crucial factors to improve the activity of HDS catalyst are higher sulfurization degree and higher aspect ratios of the active phase crystal morphology. In addition, moderate acidity and proper pores structure could be considered the key factors of deep desulfurization.

ACKNOWLEDGMENTS

We are grateful to many of our colleagues for stimulating discussions on desulfurization and hydrogenation. We would like to acknowledge the financial supports from Science and Technology Innovation Project of Shanxi Province in China (2011KTZB03-03-01) and Major Scientific and Technological Innovation Project of Shaanxi in China (2008zkc03205).

REFERENCES

1.   Y. Zhang and L. F. Zhao, "Study on Hydro Catalysis of Middle/Low Temperature Coal Tar to Clean Fuel," Coal Conversion, Vol. 32, No. 3, 2009, pp. 49-51.

2.   B. Q. Ma, P. J. Ren, Z. B. Yang and S. K. Wang, "Preparation of Fuel Oil from Coal Tar," Chemical Industry Press, Beijing, 2011.

3.   B. Pawelec, R. M. Navarro, J. M. Campos-Martin and J. L. G. Fierro, "Towards near Zero-Sulfur Liquid Fuels: A Perspective Review," Catalysis Science & Technology, No. 1, 2011, pp. 23-42. doi:10.1039/c0cy00049c

4.   L. Lan, S. H. Ge, K. H. Liu, Y. D. Hong and X. J. Bao, "Synthesis of Ni_2P Promoted Trimetallic Ni MoW/ -Al_2O_3 Catalysts for Diesel Oil Hydrotreatment," Journal of Natural Gas Chemistry, Vol. 20, No. 2, pp. 117-122.

5.   L. Zhang, X. Y. Long, D. D. Li and X. D. Gao, "Study on High-Performance Unsupported Ni-Mo-W Hydrotreating Catalyst," Catalysis Communications, Vol. 12, No. 11, 2011, pp. 927-931.

6.   Y. Zhao, M. W. Xue, M. H. Cao and J. Y. Shen, "A Highly Loaded and Dispersed Ni_2P/SiO_2 Catalyst for the Hydrotreating Reactions," Applied Catalysis B: Environmental, Vol. 104, No. 3-4, 2011, pp. 229-233.

7.   C. Sepulveda, V. Bellière, D. Laurenti, N. Escalona and R. García, "Supported Rhenium Sulfide Catalysts in Thiophene and 4,6-Dimethyldibenzo Thiophene Hydrodesulfurization: Effect of Acidity of the Support over Activities," Applied Catalysis A: General, Vol. 393, No. 1-2, 2011, pp. 288-293.

8.   A. E. Coumans, D. G. Poduval, J. A. Rob van Veen and E. J. M. Hensen, "The Nature of the Sulfur Tolerance of aAmorphous Silica-Alumina Supported NiMo(W) Sulfide and Pt Hydrogenation Catalysts," Applied Catalysis A: General, Vol. 411-412, 2012, pp. 51-59.

9.   V. L. Parola, M. L. Testa and A. M. Venezia, "Pd and PdAu Catalysts Supported over 3-MPTES Grafted HMS Used in the HDS of Thiophene," Applied Catalsis B: Environmental, Vol. 119-120, 2012, pp. 248-255. doi:10.1016/j.apcatb.2012.03.007

10.   A. Infantes-Molina, J. A. Cecilia, B. Pawelec, J. L. G. Fierro and E. Rodríguez-Castellón, "Ni_2P and CoP Catalysts Prepared from Phosphite-Type Precursors for HDSHDN Competitive

Reactions," Applied Catalysis A: General, Vol. 390, No. 1-2, 2010, pp. 253-263. doi:10.1016/j.apcata.2010.10.019

11. P. E. Boahene, K. K. Soni, A. K. Dalai and J. Adjaye, "Application of Different Pore Diameter SBA-15 Supports for Heavy Gas Oil Hydrotreatment Using FeW Catalyst", Applied Catalysis A: General, Vol. 402, No. 1-2, 2011, pp. 31-40.

12. S. Chouzier, T. Czeri, M. Roy-Auberger, C. Pichon and C. Geantet, "Decomposition of Molybdate-Hexamethylenetetramine Complex: One Single Source Route for Different Catalytic Materials," Journal of Solid State Chemistry, Vol. 184, No. 10, 2011, pp. 2668-2677.

13. Q. Ma, P. H. Zeng, S. F. Ji, H. Liu and C. Y. Li, "Effect of Metal Promoters on the Structure and Performance of the Ni_2P/SBA-15 Catalyst for Hydrodesulfurization," Acta Petrolei Sinica (Petroleum Processing Section), Vol. 27, No. 2, 2011, pp. 175-180.

14. P. F. Zhao, S. F. Ji, N. Wei, Q. Ma and H. Liu, "Effect of Boron Promoter on the Structure and Hydrodesulfurization Activity of Ni2P/SBA-15 Catalysts," Acta PhysicoChimica Sinica, Vol. 27, No. 7, 2011, pp. 1737-1742.

15. N. Wei, S. F. Ji, Y. M. Guan, H. Liu and C. Y. Li, "Influence of W on Structure and Hydrodesulfurization Performance of $W-Ni_2P/SBA$-15 Catalysts," Acta Petrolei Sinica (Petroleum Processing Section), Vol. 27, No. 6, 2011, pp. 852-858.

16. Y. N. Guo, P. H. Zeng, S. F. Ji and N. Wei, "Effect of Mo Promoter Contenton Performance of $Mo-Ni_2P/SBA$-15/ Cordierite Monolithic Catalyst for Hydrodesulfurization," Chinese Journal of Catalysis, Vol. 31, No. 3, 2010, pp. 329-334.

17. M. X. Qin, B. Yu, J. Yang and H. J. Li, "Effects of Metal Modification on Selectivity of Co-Mo/ $-Al_2O_3$ Hydrodesulifcation Catalyst," Industrial Catalysis, Vol. 17, No. 5, 2009, pp. 45-49.

18. Y. S. Zhou, Q. Wei, T. Zhang and X. J. Tao, "Effect of Phosphoric Modification on the Heavy Oil Hydrotreating Performance

of NiW/CTS Catalysts," Journal of Fuel Chemistry and Technology, Vo1. 39, No. 10, 2011, pp. 766-770.

19. D. W. Hui, Q. H. Yang, S. L. Sun and C. F. Niu, "Effect of Phosphorus on the Performance and Active Component Structure of MoCoNi/Al$_2$O$_3$ Catalyst," Petroleum Processing and Petrochemicals, Vol. 42, No. 5, 2011, pp. 1-4.

20. S. J. Ardakani and K. J. Smith, "A Comparative Study of Ring Opening of Naphthalene, Tetralin and Decalin over Mo2C/HY and Pd/HY Catalysts," Applied Catalysis A: General, Vol. 403, No. 1-2, 2011, pp. 36-47. doi:10.1016/j.apcata.2011.06.013

21. X. B. Liu, S. J. Ardakani and K. J. Smith, "The Effect of Mg and K Addition to a Mo2C/HY Catalyst for the Hydrogenation and Ring Opening of Naphthalene," Catalysis Communications, Vol. 12, No. 6, 2011, pp. 454-458.

22. S. K. Maity, M. Lemus and J. Ancheyta, "Effect of Preparation Methods and Content of Boron on Hydrotreating Catalytic Activity," Energy & Fuels, Vol. 25, No. 7, 2011, pp. 3100-3107.

23. H. L. Yin, T. N. Zhou, Y. Q. Liu, Y. M. Chai and C. G. Liu, "NiMo/Al$_2$O$_3$ Catalyst Containing Nanosized Zeolite Y for Deep Hydride Sulfurization and Hydrodenitrogenation of Diesel," Journal of Natural Gas Chemistry, Vol. 20, No. 4, 2011, pp. 441-448. doi:10.1016/S1003-9953(10)60204-6

24. Q. Wei, Y. S. Zhou, X. J. Tao, T. Zhang and Y. D. Liu, "Catalytic Activities of PY Zeolite Supported Hydrotreating Catalyst," Acta Petrolei Sinica (Petroleum Processing Section), Vol. 28, No. 1, 2012, pp. 15-20.

25. T. Klimova, O. Gutiérrez, L. Lizama and J. Amezcua, "Advantages of ZrO$_2$- and TiO$_2$-SBA-15 Mesostructured Supports for Hydride Sulfurization Catalysts over Pure TiO$_2$, ZrO$_2$ and SBA-15," Microporous and Mesoporous Materials, Vol. 133, 2010, pp. 91-99.

26. O. Y. Gutiérrez and T. Klimova, "Effect of the Support on the High Activity of the (Ni)Mo/ZrO$_2$ -SBA-15 Catalyst in the

Simultaneous Hydridesulfurization of DBT and 4,6-DMDBT," Journal of Catalysis, Vol. 281, 2011, pp. 50-62.

27.   R. Nava, A. Infantes-Molina, P. Castao, R. Guil-López and B. Pawelec, "Inhibition of CoMo/HMS Catalyst Deactivation in the HDS of 4,6-DMDBT by Support Modification with Phosphate," Fuel, Vol. 90, No. 8, 2011, pp. 2726-2737. doi:10.1016/j.fuel.2011.03.049

28.   Y. Li, D. H. Pan, C. Z. Yu, Y. Fan and X. J. Bao, "Synthesis and Hydrodesulfurization Properties of NiW Catalyst Supported on High-Aluminum-Content, Highly Ordered, and Hydrothermally Stable Al-SBA-15," Journal of Catalysis, Vol. 286, 2012, pp. 124-136.doi:10.1016/j.jcat.2011.10.023

29.   N. Malicki, G. Mali, A. A. Quoineaud, P. Bourges and L. J. Simon, "Aluminium Triplets in Dealuminated Zeolites Detected by ^{27}Al NMR Correlation Spectroscopy," Microporous and Mesoporous Materials, Vol. 129, No. 1-2, 2010, pp. 100-105.

30.   M. Nakamura, "Nanometer Catalyst Technology to Reduce the Dosage of Noble Metal," Foreign Internal Combustion Engine, Vol. 3, 2011, pp. 35-39.

31.   R. K. Sharma, B. S. Rana, D. Varma, R. Kumar and R. Tiwari, "3-D Mesoporous Titanosilicate Support for Highly Effective Hydrodesulfurization Catalysts," Microporous and Mesoporous Materials, Vol. 155, 2012, pp. 177-185.

32.   D. Valencia and T. Klimova, "Effect of the Support Composition on the Characteristics of NiMo and CoMo/(Zr)SBA-15 Catalysts and Their Performance in Deep Hydrodesulfurization," Catalysis Today, Vol. 166, 2011, pp. 91-101.

33.   Y. Kanda, C. Temma, K. Nakata, T. Kobayashi and M. Sugioka, "Preparation and Performance of Noble Metal Phosphides Supported on Silica as New Hydrodesulfurization Catalysts," Applied Catalysis A: General, Vol. 386, No. 1-2, 2010, pp. 171-178.doi:10.1016/j.apcata.2010.07.045

34.   S. W. Gong, L. J. Liu, H. F. He and Q. X. Cui, "Dibenzothiophene Hydrodesulfurization over MoP/SiO$_2$ Catalyst Prepared with

Sol-Gel Method," Korean Journal of Chemical Engineering, Vol. 27, No. 5, 2010, pp. 1419- 1422. doi:10.1007/s11814-010-0234-3

35.   T. Kan, H. Y. Wang, H. X. He, C. S. Li and S. J. Zhang, "Experimental Study on Two-Stage Catalytic Hydroprocessing of Middle-Temperature Coal Tar to Clean Liquid Fuels," Fuel, Vol. 90, No. 11, 2011, pp. 3404-3409. doi:10.1016/j.fuel.2011.06.012

36.   Z. H. Yang, L. C. Li, Y. F. Wang, J. L. Liu and X. Feng, "Preparation of Nickel Phosphide/Mesoprous-TiO_2 Catalyst and Its Hydrodesulfurization Performance," Chinese Journal of Catalysis, Vol. 33, No. 3, 2012, pp. 508-517.

37.   D. Q. Zhang, A. J. Duan, Z. Zhao, X. Q. Wang and G. Y. Jiang, "Synthesis, Characterization and Catalytic Performance of Meso-Microporous Material Beta-SBA-15- Supported NiMo Catalysts for Hydrodesulfurization of Dibenzothiophene," Catalysis Today, Vol. 175, No. 1, 2011, pp. 477-484. doi:10.1016/j.cattod.2011.03.060

38.   J. Francis, E. Guillon, N. Bats, C. Pichon and A. Corma, "Design of Improved Hydrocracking Catalysts by Increasing the Proximity between Acid and Metallic Sites," Applied Catalysis A: General, Vol. 409-410, 2011, pp. 140-147.

39.   G. Q. Cui, J. F. Wang, H. F. Fan, X. Y. Sun and Y. Jiang, "Towards Understanding the Microstructures and Hydrocracking Performance of Sulfided Ni-W Catalysts: Effect of Metal Loading," Fuel Processing Technology, Vol. 92, No. 12, 2011, pp. 2320-2327.doi:10.1016/j.fuproc.2011.07.020

40.   C. Leyva, J. Ancheyta, A. Travert, F. Maugé and L. Mariey, "Activity and Surface Properties of $NiMo/SiO_2$- Al_2O_3 Catalysts for Hydroprocessing of Heavy Oils," Applied Catalysis A: General, Vol. 425-426, 2012, pp. 1-12.

41.   L. B. Pierella, C. Saux, S. C. Caglieri, H. R. Bertorello and P. G. Bercoff, "Catalytic Activity and Magnetic Properties of Co-ZSM-5 Zeolites Prepared by Different Methods," Applied Catalysis A: General, Vol. 347, No. 1, 2008, pp. 55-61.

42. S. G. A. Ferraz, F. M. Z. Zotin, L. R. R. Araujo and J. L. Zotin, "Influence of Support Acidity of NiMoS Catalysts in the Activity for Hydrogenation and Hydrocracking of Tetralin," Applied Catalysis A: General, Vol. 384, No. 1-2, 2010, pp. 51-57.

43. Y. J. Wang, H. J. Zhang, B. Sun and X. G. Tian, "Research Progress of WO_3/ZrO_2 in Alkane Isomerization," Acta Petrolei Sinica (Petroleum Processing Section), Vol. 25, No. 2, 2009, pp. 283-290.

44. M. X. Du, Zh. F. Qin, H. Ge, X. K. Li and Z. J. Lü, "Enhancement of $Pd-Pt/Al_2O_3$ Catalyst Performance in Naphthalene Hydrogenation by Mixing Different Molecular sSieves in the Support," Fuel Processing Technology, Vol. 91, No. 11, 2010, pp. 1655-1661.doi:10.1016/j.fuproc.2010.07.001

45. L. Dimitrov, R. Palcheva, A. Spojakina and K. Jiratova, "Synthesis and Characterization of W-SBA-15 and WHMS as Supports for HDS," Journal of Porous Materials, Vol. 18, No. 4, 2011, pp. 425-434.

46. H. Shimada, K. Sato, K. Honna, T. Enomoto and N. Ohshio, "Design and Development of Ti-Modified ZeoliteBased Catalyst for Hydrocracking Heavy Petroleum," Catalysis Today, Vol. 141, No. 1-2, 2009, pp. 43-51. doi:10.1016/j.cattod.2008.04.012

47. A. Nogueira, R. Znaiguia, D. Uzio, P. Afanasiev and G. Berhault, "Curved Nanostructures of Unsupported and Al_2O_3-Supported MoS_2 Catalysts: Synthesis and HDS Catalytic Properties," Applied Catalysis A: General, Vol. 429-430, 2012, pp. 92-105.

48. V. L. Parola, B. Dragoi, A. Ungureanu, E. Dumitriu and A. M. Venezia, "New HDS Catalysts Based on Thiol Functionalized Mesoporous Silica Supports," Applied Catalysis A: General, Vol. 386, No. 1-2, 2010, pp. 43-50.

49. G. L. Yu, Y. S. Zhou, Q. Wei, X. J. Tao and Q.Y. Cui, "A Novel Method for Preparing Well Dispersed and Highly Sulfided NiW Hydridenitrogenation Catalyst," Catalysis Communications,

Vol. 23, 2012, pp. 48-53. doi:10.1016/j.catcom.2012.03.002

50. M. Ke, X. Y. Wang, Z. Q. Zhang, Z. Z. Song and Q. Z. Jiang, "Preparations and Properties of Selective Hydrodesulfurization Catalysts for FCC Gasoline," Petrochemical Technology, Vol. 40, No. 2, 2011, pp. 133-139.

51. C. Q. Li, J. W. Li, X. K. Shang, H. Wang and G. D. Sun, "Dibenzothiophene Hydrodesulfurization Performance of WP/MCM-41 Catalysts Containing Cobalt," Acta Petrolei Sinica (Petroleum Processing Section), Vol. 27, No. 6, 2011, pp. 859-865.

52. M. A. Lélias, P. J. Kooyman, L. Mariey, L. Oliviero and A. Travert, "Effect of NTA Addition on the Structure and Activity of the Active Phase of Cobalt-Molybdenum Sulfide Hydrotreating Catalysts," Journal of Catalysis, Vol. 267, No. 1, 2009, pp. 14-23.doi:10.1016/j.jcat.2009.07.006

53. M. A. Lélias, E. Le. Guludec, L. Mariey, J. van Gestel and A. Travert, "Effect of EDTA Addition on the Structure and Activity of the Active Phase of Cobalt–Molybdenum Sulfide Hydrotreatment Catalysts," Catalysis Today, Vol. 150, No. 3-4, 2010, pp. 179-185.doi:10.1016/j.cattod.2009.07.107

54. K. K. Soni, K. C. Mouli, A.K. Dalai and J. Adjaye, "Effect of Ti Loading on the HDS and HDN Activity of KLGO on NiMo/ TiSBA-15 Catalysts," Microporous and Mesoporous Materials, Vol. 152, 2012, pp. 224-234.

55. K. C. Mouli, K. Soni, A. Dalai and J. Adjaye, "Effect of Pore Diameter of Ni-Mo/Al-SBA-15 Catalysts on Thehydrotreating of Heavy Gas Oil," Applied Catalysis A: General, Vol. 404, No. 1-2, 2011, pp. 21-29. doi:10.1016/j.apcata.2011.07.001

56. P. Biswas, P. Narayanasarma, C. M. Kotikalapudi, A. K. Dalai and J. Adjaye, "Characterization and Activity of ZrO_2 Doped SBA-15 Supported NiMo Catalysts for HDS and HDN of Bitumen Derived Heavy Gas Oil," Industrial & Engineering Chemistry Research, Vol. 50, No. 13, 2011, pp. 7882-7895.

57. M. A. Al-Daous and S. A. Ali, "Deep Desulfurization of Gas Oil over NiMo Catalysts Supported on Alumina-Zirconia Composites," Fuel, Vol. 97, 2012, pp. 662-669.

58. T. Kimura, K. Sakashita, X. H. Li and S. Asaoka, "Catalytic Roles of Nano-Sized Oxides Composed with Zeolite for Hydrocracking, Catalytic Cracking and Reforming," Catalysis Surveys from Asia, Vol. 15, 2011, pp. 259-266.

59. C. Li, W. A. Deng, X. W. Li, B. Q. Mu and S. F. Li, "Pilot Tests of Medium/Low Temperature Coal Tar Oil Hydro-Upgrading with Heavy Oil Hydro-Treating Catalysts," Petroleum Refinery Engineering, Vol. 41, No. 9, 2011, pp. 32-35.

Coating of Medical-Grade PVC Material with ZnO for Antibacterial Application

Huaxiang Lin[1], Luyao Ding[1], Weihua Deng[1], Xuxu Wang[1], Jinlin Long[1], and Qun Lin[2]

[1]Research Institute of Photocatalysis, Fujian Provincial Key Laboratory of Photocatalysis, State Key Laboratory Breeding Base, Fuzhou University, Fuzhou, China
[2]First Affiliated Hospital, Fujian Medical University, Fuzhou, China

ABSTRACT

The ZnO sol well-crystallized was prepared by the sol-gel method. The ZnO films were coated on medical-grade PVC surface by the improved organic-inorganic interfacial adhesion method.

The physical and photocatalytic properties of the samples were characterized by XRD, SEM, DRS spectra and measured by the photodegradation reaction of Rhodamine B (RhB) and anti-bacteria for Escherichia coli (E. coli), respectively. The results show that pretreatment of PVC by the mix solution of THF-PVC helps to improve the amount and adhesion strength of ZnO suspension to PVC surface. The photocatalytic and antibacterial properties of the THF-ZnO/PVC film are better than that of the ZnO/PVC and neat PVC. Under UV irradiation, the THF-ZnO/PVC film shows the best antibacterial properties with 99% kill rate of bacteria.

INTRODUCTION

The increasing use of polymer materials such as polyethylene, polyurethanes, and poly (vinyl chloride) (PVC) in the hospital care has led to a concomitant increase in the incidence of biomaterial-related infections (BRI) [1]. Adhesion of bacteria to biomaterials led to the formation of biofilm on the surface, which plays a crucial role in the pathogenesis of the BRI [1-3]. The growth and production of biofilm protect the bacteria from the host defence mechanisms and external agents as the drug treatments [1,2,4-7], which makes the cure of the bacterial infections quite difficult and requires either higher doses or more potent antibiotics.

In order to efficiently prevent or reduce biofilm formation, many efforts have been done to enhance the antibacterial properties of biomaterials. Some efforts such as modifying the physicochemical properties of biomaterial surface, coating with silver, azidation treatment, antibiotic impregnation into the polymer matrix, have been examined in recent years [8-12].

Recently, coating semiconductor photocatalyst on materials' surface for antibacterial activity has attracted great interest due to the following advantages when compared with other methods: 1) excellent photocatalytic and antimicrobial activity, 2) nontoxicity and bio-compatibility, 3) strong physicochemical stability and durable antibacterial properties. The antibacterial mechanism of semiconductor photocatalysis is mainly based on its UV photoactive property. Taking ZnO as an example was shown in Equations (1)-(4) [13].

$$ZnO + hv \rightarrow e^- + h^+ \tag{1}$$

$$h^+ + OH^- \rightarrow \bullet OH \tag{2}$$

$$e^- + O_2 \rightarrow O_2 \bullet \tag{3}$$

$$O_2 \bullet + 2H^+ + e^- \rightarrow H_2O_2 \tag{4}$$

When ZnO was exposed to UV light with the wavelength less than 388 nm, the semiconductor surface generated valence band holes and conduction band electrons as in Equation (1). The holes could react with water or surface hydroxyls (H_2O/ OH^-) to generate hydroxyl radicals (HO•), while conduction band electrons reacted with adsorbed molecular oxygen (O_2) to generate superoxide radical (O_2•). The radicals (O_2• and HO•) can destroy and shrink bacterial cells by reacting with the organism of bacterial cells in the photocatalytic process [14, 15]. Among various oxides semiconductor photocatalysis, TiO_2 and ZnO are generally considered to be the most suitable for antibacterial agents [15-18]. TiO_2 film coated on different materials' surface is effective at killing bacteria and viruses under UV irradiation [15,16]. ZnO as antibacterial agents has been reported exhibiting higher antibacterial activity than other metal oxide nanoparticles [19]. It has supposed that ZnO not only kills bacteria under UV irradiation bacteria but also inhibits the bacterial growth under normal visible lighting conditions [20]. To extend the usage of antibacterial agents, many studies have been done to coat ZnO film on many materials' surface that was widely used in life such as glass, ceramic, stainless steel, polymer, and so on. In general, to obtain highly adherent and antibacterial activity of ZnO film, heating the materials that have been modified by ZnO film at high temperature is required. However, one of the main barriers to coat ZnO on polymer is the low heat resistance of polymer. Low heat resistance of polymer disables the routine coating method to be used to modify inorganic ZnO film on polymer.

Recently, organic-inorganic composite membrane by mixing inorganic and organic materials has been supposed to use because of its simplicity, mild conditions and stability [21,22]. In this work, a similar organic-inorganic interfacial adhesion method was used to coat ZnO film on polymer materials. First, suspension of well-crystallized ZnO was prepared by sol-gel at low temperature. Then, to improve the adhesion strength of ZnO to PVC, the mix solution of PVC and THF was used to pretreat the surface of PVC sheet. After pretreated by the mix solution of PVC and THF, PVC sheets were coated with ZnO film by dip-coating with suspension ZnO. The photocatalytic and antibacterial activity of ZnO film was investigated. To evaluate the antibacterial activity after modified by ZnO, the performance of ZnO film was compared with neat PVC under similar operating conditions.

MATERIALS AND METHODS

Material

Medical-grade PVC sheets were purchased from Tyco Healthcare International Trading (Shanghai) Co., Ltd., and were washed with absolute alcohol before use. Escherichia coli were cultured at 37°C in Luria-Bertani (LB) medium (pH 7.0) containing tryptone 1%, yeast extract 0.5%, sodium chloride 1%. When tested, the bacterial concentration was adjusted to the required concentration (10^4 - 10^5 cfu/ml). Other reagents were of analytical grade and used as received.

Preparation of ZnO Sol

$Zn(Ac)_2 \times 2H_2O$ was dissolved in methanol to obtained a 0.075 M solution of Zn (Ac). Ammonia was dropped into the solution to adjust the pH to 8 and then stirred the solution at 65°C for 2 h to obtain ZnO sol.

Coating ZnO on PVC Surface

The ZnO film was prepared by dip-coating method with ZnO suspension as precursor. The PVC sheets (1.5 × 4.5 cm) were pre-immersed in THF-PVC solution for 10s. Immediately, the PVC samples were dipped into ZnO colloidal sol and were slowly pulled out of the sol at a speed of 1200 mm/h in an ambient atmosphere. The ZnO gel film on PVC were dried in an oven at 60°C for 30 min and then were dried at 120°C for 2 h. After seven such coating steps, the transparent ZnO film on PVC were obtained. The PVC sheets without pretreatment were coated with ZnO under similar operating conditions for comparison. For ease of presentation, the ZnO film with pretreatment were labelled as THF-ZnO/PVC and that without pretreatment were labelled as ZnO/PVC.

Structure and Morphology Characterizations

The crystal structure of the sample was characterized by X-ray diffraction (XRD) (Bruker D8 Advance powder diffractometer). The scanning electron microscopy (SEM) was performed on a Philips XL30 ESEM system. The UV-Vis spectra were obtained in the diffuse reflection mode using a Carry 500 spectrometer at a resolution of 2 nm. The spectra of the PVC, ZnO/PVC and THF-ZnO/ PVC sheets were recorded in the ultraviolet and visible in the wavelength range from 250 to 800 nm.

Photocatalytic and Antibacterial Property

Photocatalytic Activity

The photocatalytic activities of PVC, ZnO/PVC and THF-ZnO/ PVC were determined by monitoring the degradation of organic dye Rhodamine B (RhB). A quartz tube reactor with condenser pipe was used and four 4W-UV lamps with wavelength at 365 nm were fixed around the reactor. The PVC sheets (1.5 × 4.5 cm) were immersed in 20 ml aqueous solution of RhB (1 × 10^{-5} mol/dm^3) and kept for 12 h in black prior to irradiation. The concentration of RhB at different irradiation times was monitored by Varian UV-260 spectrometer.

Bacterial Adhension

The neat PVC, ZnO/PVC and THF-ZnO/PVC sheets were immersed in the aqueous solution of Escherichia coli (E. coli), and kept at 37°C for 24 h. The sheets were taken out and rinsed gently with sterile phosphate buffered saline (PBS) to remove the non-adherent bacteria. Then, the bacteria adhered on the sheets were eluted into 5 ml of sterile PBS in an ultrasonic cleaner for 5 min. The number of the eluted bacteria was then determined by colony counts (CFU). The adherent number was expressed by the ratio of the total adherent bacterial to the area of the measured sample.

Antibacterial Property

ZnO/PVC sheets (2.5 × 2.5 cm) were placed on sterile plates, and then 0.5 ml broth inoculated with 10^4 cfu/ml of E. coli was added onto the surfaces. The samples were irradiated with an 8 W UV lamp (with wavelength at 365 nm) for 150 min. After irradiated, the sheets were rinsed with sterile phosphate buffered saline (PBS) and the number of viable bacteria was determined by colony counts (CFU). The blank PVC sample was also test for comparison. The antibacterial property of PVC, ZnO/ PVC and THF-ZnO/PVC was valued by the kill percentage, a ratio of the dead number of the bacterial to the initial number of the bacterial.

RESULTS AND DISCUSSION

Structure and Morphology

The ZnO nanoparticles were prepared by simple sol-gel method. The size analysis of ZnO sol indicated that most of ZnO colloid particles are located in the range of 50 - 70 nm (Figure 1). ZnO colloid particles were dried into powder at 120°C for X-ray diffraction (XRD) analysis. Several obvious peaks are observed at 2q values at 31.6, 34.3, 36.1, 47.3, 56.4, 62.5 and 67.7, which correspond to wurtzite structure of hexagonal ZnO [23] (Figure 2). No other phases are observed in the patterns, indicating that the sample that prepared by this method are pure wurtzite ZnO.

Figure 3 shows the SEM images of neat PVC, ZnO/PVC and THF-ZnO/PVC. The neat PVC surface is smooth with little flaws. After coated only with ZnO colloid, the morphology of ZnO/PVC become rougher and a small ZnO nanostructure sheet spread on the ZnO/PVC surface. However, when PVC were pretreated by THF-PVC solution and subsequently coated with ZnO colloid, many ZnO nanostructure sheets stacked on THFZnO/PVC surface closely and the surface of THF-ZnO/ PVC was composed by irregularity cross-linked ZnO sheets. This result indicated that pretreated PVC by THFPVC solution improves the adherent amount of ZnO on PVC surface. The Optical absorption property of PVC, ZnO/PVC and THF-ZnO/PVC were examined with UVVis reflectance spectra (DRS), as shown in Figure 4. The DRS spectra show that ZnO/PVC and THF-ZnO/PVC exhibit significant enhancement of light absorption in the UV region than nude PVC, which is attributed to bandgap excitation of ZnO. Another difference in the DRS spectra is that the absorption baseline of the resulting samples in visible light region is decrease in the order of PVC > ZnO/PVC > THF-ZnO/PVC. This is originnated from the transparent nature of the samples. ZnO deposited on PVC reduces transparency of PVC, which causes more of incident light reflect and consequently enhances the baseline absorption in visible light region. The more amount of ZnO coated on PVC, the lower baseline absorption of the sample.

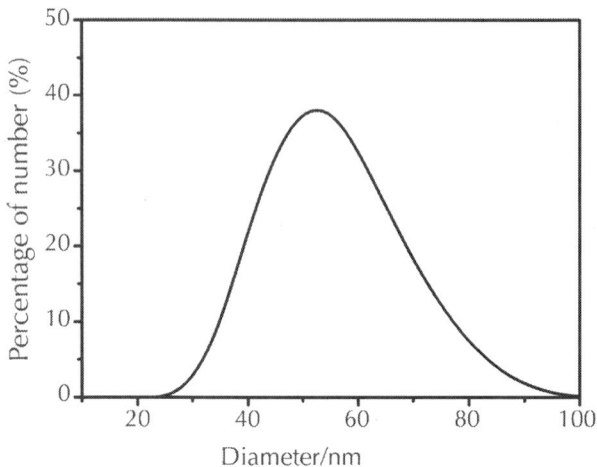

Figure 1: Size distribution of ZnO colloid particles.

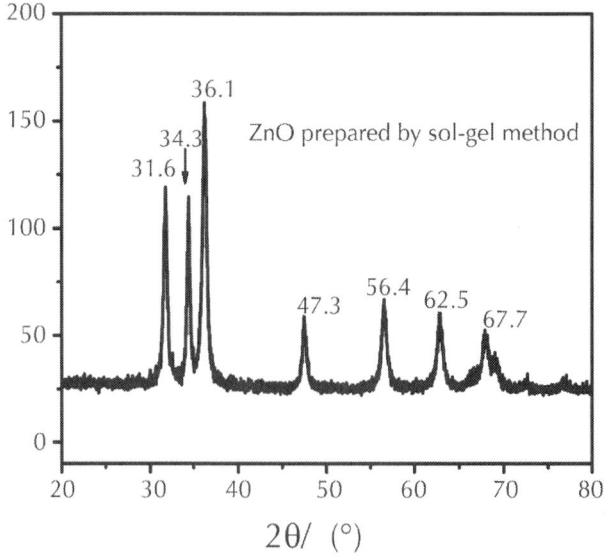

Figure 2: XRD spectra of ZnO.

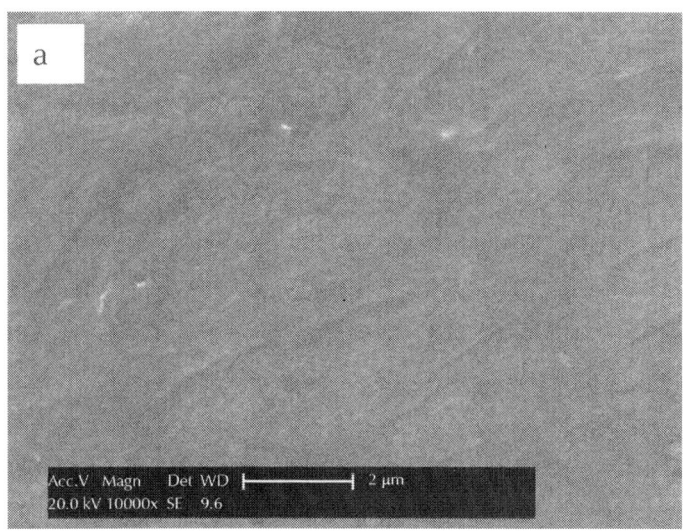

(a)

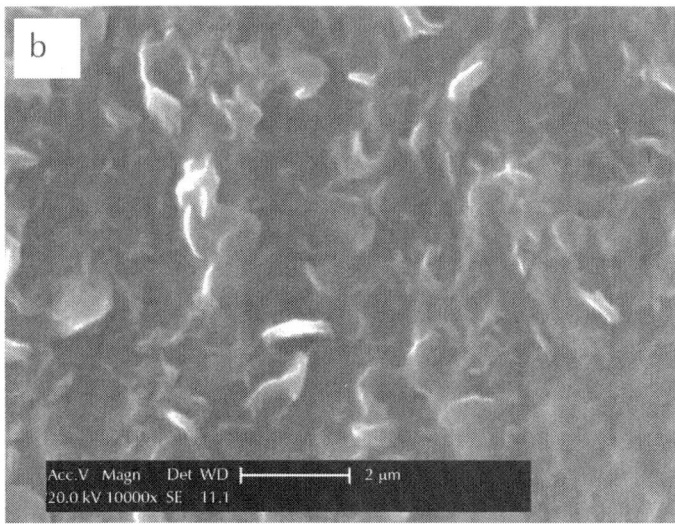

(b)

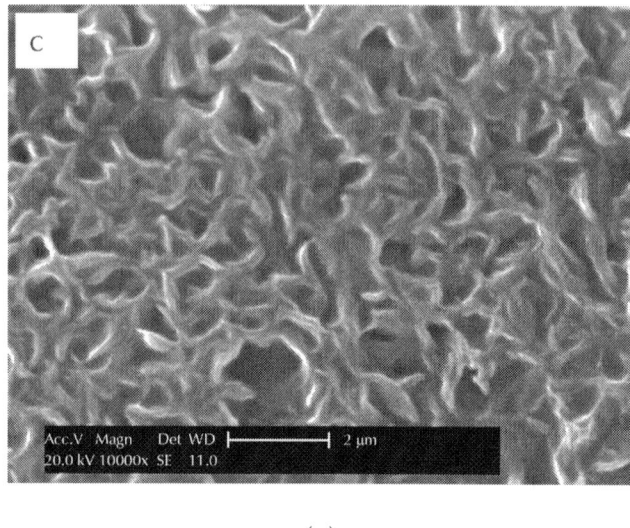

(c)

Figure 3: SEM images of different PVC surface; a) neat PVC; b) ZnO/PVC; THF-ZnO/PVC.

Photocatalytic Activity

Figure 5 shows the change in photocatalytic conversions of RhB with reaction time on the nude PVC, ZnO/PVC and THF-ZnO/PVC. As can been seen from these spectra, the nude PVC sheet exhibits no photocatalytic activity. Only about 25% of RhB are photodecomposed following about 300 min of illumination. The photocatalytic activity of ZnO/PVC samples exhibit a little better than that of neat PVC, with the conversion rate of RhB increases to about 50% of RhB upon about 300 min of illumination. The pretreatment of THF-PVC solution for PVC has much great effect on the photocatalytic activity of the resulting sample. As can be seen form spectra c in Figure 5, the degraded rate of RhB on THF-ZnO/PVC is greatly increased and about 90% of RhB are degraded upon about 300 min of illumination. The result shows that pretreatment of THF-PVC for PVC is beneficial for the coating of ZnO on PVC surface.

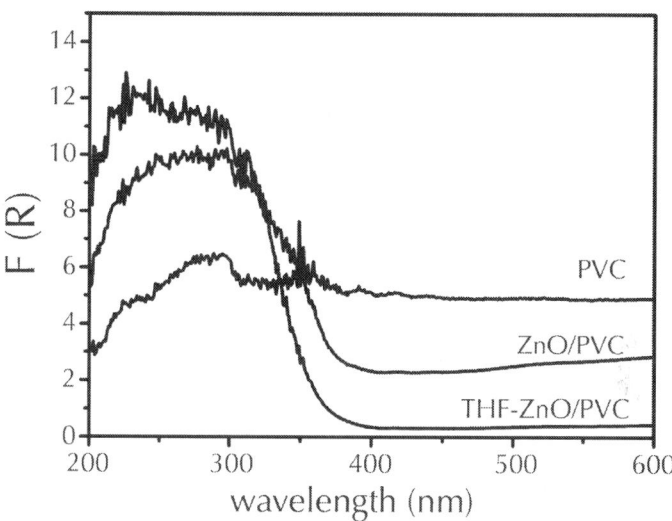

Figure 4: UV-vis diffuse reflectance spectra of PVC, ZnO /PVC and THF-ZnO/PVC.

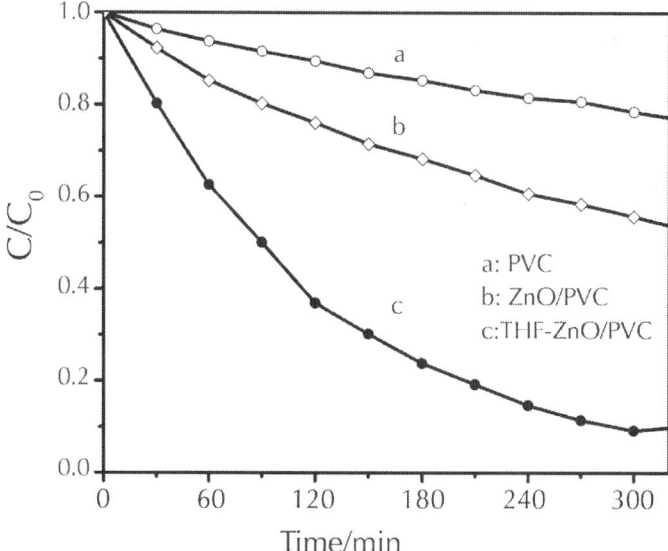

Figure 5: Photocatalytic degradation conversion of RhB on PVC, ZnO/PVC and THF-ZnO/PVC.

Bacterial Adhesion and Sterilization Behavior

Figure 6 shows the adherent number of E. coli to the surface of nude PVC, ZnO/PVC and THF-ZnO/PVC under dark condition. After culture bacteria on the sample surface in dark for 18h, the number of E. coli adhered on nude PVC is similar to initial number of E. coli (Figure 6(d)), showing that the nude PVC material is inertial to prevent adhesion of bacterial. The adherent number of E. coli on ZnO/PVC decreases a little when compared with that of nude PVC as shown in Figures 6(a) and (b), suggesting that coating ZnO on PVC helps to enhance antibacterial activity of PVC. Especially, when the PVC was pretreated with mix solution of THF-PVC and subsequently coated with ZnO, the antibacterial activity of THF-ZnO/PVC increases greatly with the adherent number of E. coli is almost zero (Figure 6(c)). The result suggests that THF-ZnO/PVC can effective prevent the adhesion and growth of bacterial on PVC surface. Figure 7 shows the kill percentage of E. coli on PVC, ZnO/PVC and THF-ZnO/PVC under UV irradiation for 150 min. PVC exhibits no antibacterial activity under UV light. And ZnO/PVC shows a little antibacterial activity with the survival ratio of E. coli on ZnO/PVC is 46.8% after irradiation for 150 min. However, the THFZnO/PVC exhibits stronger antibacterial property. More than 99% of the bacteria were killed under UV irradiation for 150 min.

It has been reported that it is the photocatalytic property of metal oxide that is responsible for the excellent sterilization properties [15,24]. In other words, the higher the photocatalytic property, the higher the sterilization activity, and vice versa. This proposition was identified by the photocatalytic property and sterilization activity of TiO_2 [15]. In the present work, the photocatalytic property and sterilization activity of THF-ZnO/PVC are much better than that of ZnO/PVC. The main reason is may be that the amount of ZnO on THF-ZnO/PVC is larger than that on ZnO/PVC. Under UV irradiation, the amount of electron and hole produced on THF-ZnO/PVC were higher than that on ZnO/PVC that resulted in higher concentration of radicals ($O_2\bullet$ and $HO\bullet$) on THF-ZnO/ PVC.

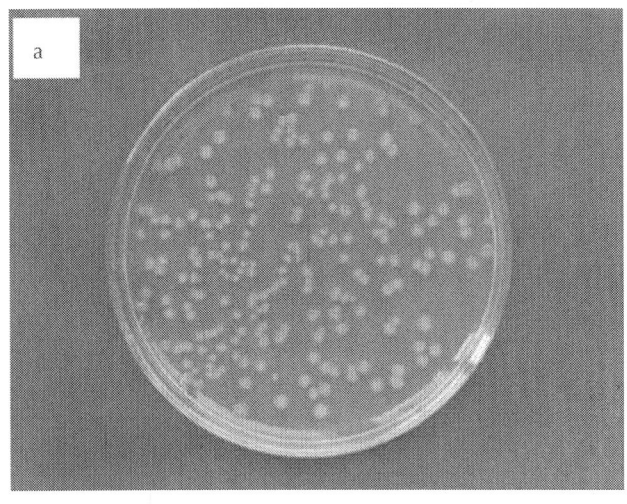

(a)

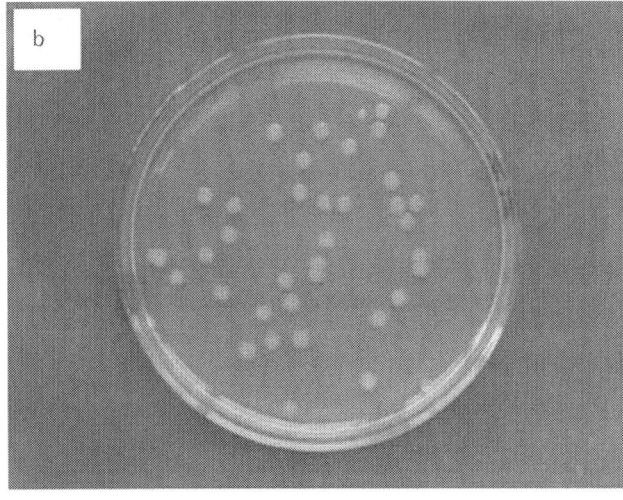

(b)

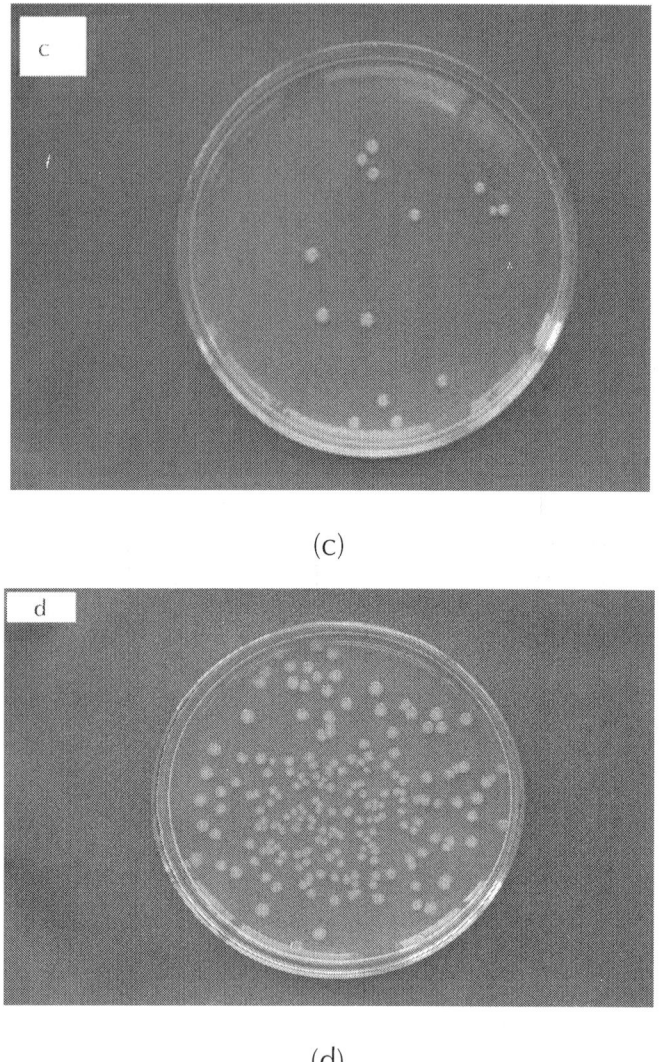

(c)

(d)

Figure 6: Bacteria adherent number on different PVC surface under dark condition. a) neat PVC; b) ZnO/PVC; THF-ZnO/PVC; d) initial number.

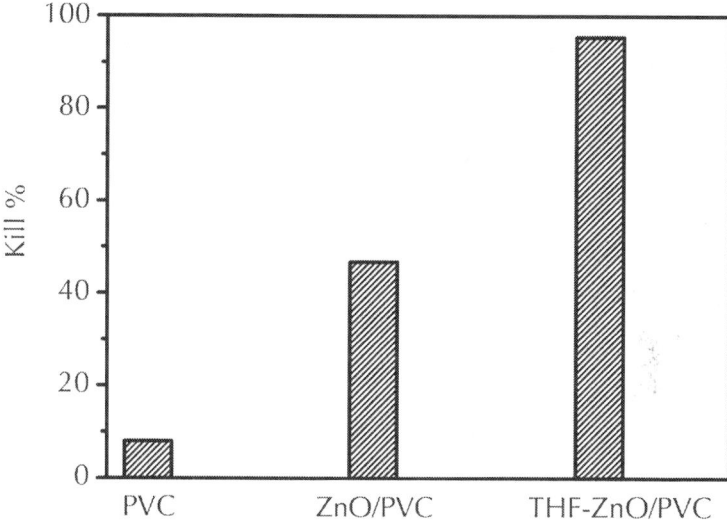

Figure 7: Kill percentage of bacteria on the samples under UV irradiation for 150 min.

CONCLUSIONS

This work infers that the antibacterial properties of medical-grade PVC material can be enhanced by coating with ZnO. After pretreatment by THF-PVC solution and subsequently coated with ZnO, the PVC surface can not only inhibit the growth of bacteria on dark conditions effectively but also kill bacteria under UV irradiation. The result provides a convenient method that can enhance the adhesive strength and the amount of inorganic oxide film coated on organic materials. It is also easy to operate and can be applied in various substrates.

ACKNOWLEDGMENTS

This work was financially supported by the NSFC (Grants Nos. 21003021, 21173044 and 21203029), the great Science and

Technology Project of Fujian Province of P. R. China (2012Y4003), the National Science Foundation for Fostering Talents in Basic Research of China (No. J1103303), and National Basic Research Program of China (973 Program, No. 2012CB722607) and the Science and Technology project of Fuzhou University (2011-XQ-09).

REFERENCES

1. S. Lakshmi, S. S. Pradeep and J. A. Kumar, "Bacterial Adhesion onto Azidated Poly (Vinyl Chloride) Surfaces," Journal of Biomedical Materials Research, Vol. 61, No. 1, 2002, pp. 26-32. http://dx.doi.org/10.1002/jbm.10046.abs

2. P. Vergidis and R. Patel, "Novel Approaches to the Diagnosis, Prevention, and Treatment of Medical Device-Associated Infection," Infectious Disease Clinics of North America, Vol. 26, No. 1, 2012, pp. 173-186. http://dx.doi.org/10.1016/j.idc.2011.09.012

3. D. S. Jones, J. G. McGovern and D. A. Woolfson, "Role of Physiological Conditions in the Oropharynx on the Adherence of Respiratory Bacterial Isolates to Endotracheal Tube Poly (Vinyl Chloride)," Biomaterials, Vol. 18, No. 6, 1997, pp. 503-510. http://dx.doi.org/10.1016/S0142-9612(96)00170-6

4. Y. H. An and J. R. Friedman, "Prevention of Sepsis in Total Joint Arthroplasty," Journal of Hospital Infection, Vol. 33, No. 2, 1996, pp. 93-108. http://dx.doi.org/10.1016/S0195-6701(96)90094-8

5. Y. H. An and J. R. Friedman, "Concise Review of Mechanisms of Bacterial Adhesion to Biomaterial Surfaces," Journal of Biomedical Materials Research, Vol. 43, No. 3, 1998, pp. 338-348. http://dx.doi.org/10.1002/(SICI)1097-4636(199823)43:3%3C338::AID-JBM16%3E3.0.CO;2-B

6. A. G. Gristina, "Biomaterial-Centered Infection: Microbial Adhesion versus Tissue Integration," Science, Vol. 237, No.

4822, 1987, pp. 1588-1595.http://dx.doi.org/10.1126/science.3629258

7. A. G. Gristina, C. D. Hobgood and L. X. Webb, "Adhesive Colonization of Biomaterials and Antibiotic Resistance," Biomaterials, Vol. 8, No. 6, 1987, pp. 423-426.http://dx.doi.org/10.1016/0142-9612(87)90077-9

8. N. P. Desai, S. F. A. Hossainy and J. A. Hubbei, "Surface-Immobilized Polyethylene Oxide for Bacterial Repellence," Biomaterials, Vol. 13, No. 7, 1992, pp. 417- 420.http://dx.doi.org/10.1016/0142-9612(92)90160-Pl

9. K. D. Park, Y. S. Kim and D. K. Hun, "Bacterial Adhesion on PEG Modified Polyurethane Surfaces," Biomaterials, Vol. 19, No. 7-9, 1998, pp. 851-859.http://dx.doi.org/10.1016/S0142-9612(97)00245-7

10. Z. Zdanowski, B. Koul and E. Hallberg, "Inflence of Heparin Coating on in Vitro Bacterial Adherence to Poly (Vinyl Chloride) Segments," Journal of Biomaterials Science, Polymer Edition, Vol. 8, No. 11, 1997, pp. 825- 832.http://dx.doi.org/10.1163/156856297X00029

11. D. J. Balazs, K. Triandafillu and Y. Chevolot, "Surface Modification of PVC Endotracheal Tubes by Oxygen Glow Discharge to Reduce Bacterial Adhesion," Surface and Interface Analysis, Vol. 35, No. 3, 2003, pp. 301-309. http://dx.doi.org/10.1002/sia.1533

12. K. Triandafillu, D. J. Balazs and B. D. Aronsson, "Adhesion of Pseudo-Monas Aeruginosa Strains to Untreated and Oxygen-Plasma Treated Poly (Vinyl Chloride) (PVC) from Endotracheal Intuba-tion Devices," Biomaterials, Vol. 24, No. 8, 2003, pp. 1507-1518.http://dx.doi.org/10.1016/S0142-9612(02)00515-X

13. T. J. Whang, M. T. Hsieh and H. H. Chen, "Visible-Light Photocatalytic Degradation of Methylene Blue with Laser-Induced Ag/ZnO Nanoparticles," Applied Surface Science, Vol. 258, No. 7, 2012, pp. 2796-2801. http://dx.doi.org/10.1016/j.apsusc.2011.10.134

14. W. K. Ho, J. C. Yu and S. C. Lee, "Photocatalytic Activity and Photo-Induced Hydrophilicity of Mesoporous TiO_2 Thin films Coated on Aluminum Substrate," Applied Catalysis B: Environmental, Vol. 73, No. 1-2, 2007, pp. 135-143.http://dx.doi.org/10.1016/j.apcatb.2006.06.019

15. H. X. Lin, Z. T. Xu and X. X. Wang, "Photocatalytic and Antibacterial Properties of Medical-Grade PVC Material Coated With TiO_2 Film," Journal of Biomedical Materials Research, Vol. 87, No. 2, 2008, pp. 425-431. http://dx.doi.org/10.1002/jbm.b.31120

16. W. Y. Su, S. H. Wang and X. X. Wang, "Plasma PreTreatment and TiO_2 Coating of PMMA for the Improvement of Antibacterial Properties," Surface and Coatings Technology, Vol. 205, No. 2, 2010, pp. 465-469. http://dx.doi.org/10.1016/j.surfcoat.2010.07.013

17. Y. Li, W. Zhang and J. F. Niu, "Mechanism of Photogenerated Reactive Oxygen Species and Correlation with the Antibacterial Properties of Engineered Metal-Oxide Nanoparticles," ACS Nano, Vol. 6, No. 6, 2012, pp. 5164- 5173.http://dx.doi.org/10.1021/nn300934k

18. A. Thill, O. Zeyons, O. Spalla and F. Chauvat, "Cytotoxicity of CeO_2 Nanoparticles for Escherichia coli. PhysicoChemical Insight of the Cytotoxicity Mechanism," Environmental Science & Technology, Vol. 40, No. 19, 2006, pp. 6151-6156. http://dx.doi.org/10.1021/es060999b

19. K. H. Tam, A. B. Djurisic, C. M. N. Chan and Y. H. Leung, "Antibacterial Activity of ZnO Nanorods Prepared by a Hydrothermal Method," Thin Solid Film, Vol. 516, No. 18, 2008, pp. 6167-6174. http://dx.doi.org/10.1016/j.tsf.2007.11.081

20. N. Jones, B. Ray, K. T. Ranjit and A. C. Manna, "Antibacterial Activity of ZnO Nanoparticle Suspensions on a Broad Spectrum of Microorganisms," FEMS Microbiology Letters, Vol. 279, No. 1, 2008, pp. 71-76. http://dx.doi.org/10.1111/j.1574-6968.2007.01012.x

21. Y. Yang, H. Zhang, P. Wang and Q. Zheng, "The Influence of Nano-Sized TiO_2 Fillers on the Morphologies and Properties of PSF UF Membrane," Journal of Membrane Science, Vol. 288, No. 1-2, 2007, pp. 231-238. http://dx.doi.org/10.1016/j.memsci.2006.11.019

22. R. A. Damodara, S. J. Youa and H. H. Chou, "Study the Self Cleaning, Antibacterial and Photocatalytic Properties of TiO_2 Entrapped PVDF Membranes," Journal of Hazardous Materials, Vol. 172, No. 2-3, 2009, pp. 1321-1328.http://dx.doi.org/10.1016/j.jhazmat.2009.07.139

23. S. W. Liu, C. Li and J. G. Yu, "Improved Visible-Light Photocatalytic Activity of Porous Carbon Self-Doped ZnO Nanosheet-Assembled Flowers," Cryst Eng Comm, Vol. 13, No. 7, 2011, pp. 2533-2541. http://dx.doi.org/10.1039/c0ce00295j

24. M. Bekbolet, "Photocatalytic Bactericidal Activity of TiO_2 in Aqueous Suspensions of E. coli," Water Science and Technology, Vol. 35, No. 11-12, 1997, pp. 95-100.http://dx.doi.org/10.1016/S0273-1223(97)00241-2

Chapter 5

Enhanced Electrochemical Properties of LiFePO$_4$ as Positive Electrode of Li-Ion Batteries for HEV Applicationt

Christian M. Julien[1], Karim Zaghib[2], Alain Mauger[3], and Henri Groult[1]

[1]Physicochimie des Electrolytes, Colloïdes et Systèmes Analytiques, Université Pierre et Marie Curie, Paris, France
[2]Institut de Recherche d'Hydro-Québec, Varennes, Canada
[3]Institut de Minéralogie et Physique de la Matière Condensée, Université Pierre et Marie Curie, Paris, France

ABSTRACT

LiFePO$_4$ materials synthesized using FePO$_4$ (H$_2$O)$_2$ and Li$_2$CO$_3$ blend were optimized in view of their use as positive electrodes in Li-ion batteries for hybrid electric vehicles. A strict control of the structural properties was made by the combination of X-ray diffraction, FT-infrared spectroscopy and magnetometry. The impact of the ferromagnetic clusters (γ-Fe_2O_3 or Fe_2P) on the electrochemical response was examined. The electrochemical performances of the optimized $LiFePO_4$ powders investigated at 60°C are excellent in terms of capacity retention (153 mAh·g^{-1} at 2C) as well as in terms of cycling life. No iron dissolution was observed after 200 charge-discharge cycles at 60°C for cells containing Li foil, $Li_4Ti_5O_{12}$, or graphite as negative electrodes.

INTRODUCTION

Since the introduction of lithium-ion batteries based on lithium cobaltate ($LiCoO_2$) by Sony in 1991, great efforts have been addressed to find an alternative material with both sides of the battery. However, the expansion of their applications from the portable market to the electric (EVs) and hybrid vehicles (HEVs) requests lower cost and better safety characteristic electrode material. Among the well-known Li-insertion compounds, the olivine $LiFePO_4$ (LFP) compound is being extensively investigated as a positive electrode material for Li-ion batteries because of its low cost, low toxicity, and relatively high theoretical specific capacity of 170 mAh·g^{-1} [1,2]. The current debate for the utilization of $LiFePO_4$ in large-size batteries (for HEV, for instance), is mainly focused on the perceived poor rate capability because of a low electronic conductivity. Another aspect concerns the material purity and the non-migration of iron ions through the electrolyte. The high-temperature performance is also a critical issue because batteries may be operated at elevated temperatures (around 60°C). The early drawback of highly resistive $LiFePO_4$ has been resolved by painting the particle surface with carbon [3-6].

Recently, significant effort has been underway to improve $LiFePO_4$ by developing a new synthesis route via carbon coating [7]. The 1D Li channels make the olivine performance sensitive not only to particle size, but also to impurities and stacking faults that block the channels. Various types of iron-based impurities have been identified in the olivine framework: for examples γ-Fe_2O_3, Fe_3O_4, $Li_3Fe_2(PO_4)_3$, $Fe_2P_2O_7$, Fe_2P, Fe_3P, $Fe_{75}P_{15}C_{10}$, etc. Critical quality control of the product is necessary to obtain a complete understanding of synthesis conditions using combination of experiments such as Raman spectroscopy and magnetic measurements [8-12].

In this paper, we report the results obtained on several samples of $LiFePO_4$ (LFP) with special attention to the new generation of phospho-olivine materials used in lithium cells operating at 60°C. The magnetic properties are correlated with the electrochemical performance of the positive electrode materials. Magnetization and susceptibility measurements appear to be a powerful probes for impurity detection at very low concentration of trivalent iron (<1 ppm). Electrochemical performances of Liion cells with $Li_4Ti_5O_{12}$ (LTO) negative electrode are reported with a strict control of iron dissolution by postmortem analysis.

EXPERIMENTAL

The optimized $LiFePO_4$ material was synthesized by solidstate reaction. Samples were prepared from $FePO_4(H_2O)_2$ and Li_2CO_3. A stoichiometric amount of precursors was thoroughly mixed together in isopanol. After drying, the blend was heated at 500°C - 800°C for 8 h under reducing atmosphere. Four samples have been considered heated at carbon-coated $LiFePO_4$ (C-LFP) was prepared with sucrose and cellulose acetate as the carbon precursors in acetone solution according to the following procedure. The carbon-free powder was mixed with the carbon precursors. The dry additive corresponded to 5 wt% carbon in $LiFePO_4$. After drying, the blend was heated at 700°C for 4 h under argon atmosphere. The quantity of carbon coat represents about 1 wt% of the material (C-detector, LECO Co., and CS 444). It should be noted that the choice of this moderate sintering temperature minimizes the amount of Fe^{3+} ions present in the powder since the presence of Fe^{3+} has been detected by Mössbauer experiments at sintering temperatures below 500°C, and both trivalent Fe_2O_3 and $Li_3Fe_2(PO_4)_3$ are formed in such large quantities that they are detected by X-rays by sintering above 800°C [13]. Nevertheless, we know from our prior work [10, 11] that $LiFePO_4$, even with an intermediate sintering temperature in the range 500°C - 800°C, does contain Fe_2O_3 nanoparticles, although in such small quantities that they can be detected only by investigation of magnetic properties. X-ray diffractometry (XRD) was carried out with a Philips X'Pert apparatus equipped with a CuKα X-ray source (λ = 1.5406 Å). Slice views were examined with a scanning electron microscope (SEM, Philips XL30). Fourier transform infrared (FTIR) absorption spectra were recorded with a Fourier transform interferometer (model Bruker IFS113v) in the wavenumber range 150 - 1400 cm^{-1} at a spectral resolution of 2 cm^{-1}. Magnetic measurements (susceptibility and magnetization) were carried out with a fully automated magnetometer (MPMS-5S from Quantum Design) using an ultra-sensitive Superconducting Quantum Interference Device (SQUID) in the temperature range 4 - 300 K. The experimental details are given elsewhere [11]. The electrochemical properties of $LiFePO_4$ were measured at 60°C in cells with metallic lithium as the negative electrode. The electrolyte was 1 M $LiPF_6$ in EC/DEC (1/1) solvent. The measurements were carried out following the experimental procedure previously described [14] using the coffee-bag technology developed at Hydro-Québec. Coffee-bag or laminated battery technology was described by Zaghib and Armand [15].

RESULTS AND DISCUSSION

Structure and Morphology of LiFePO4

Figure 1 shows the typical XRD patterns of the $LiFePO_4$ electrode material. The XRD pattern of sample synthesized from the mixture $FePO_4(H_2O)_2 + Li_2CO_3$ agrees very well with that of phospho-olivine LiFePO$_4$ [16] and no impurity was detected. The XRD diagram of the new generation of LFP after 200 cycles (47 days) at 60°C are also shown in Figure 1. There is no change in the olivine structure after cycling at 60°C. We observed Bragg lines with the same intensity as that for the pristine material. The capacity loss was below 3% in 100 cycles for this optimised electrode material, which also displays excellent capacity retention. First, we present an overview of the high-temperature performance for an optimised LFP sample. The coffee-bag cell was charged and discharged at C/8 for the first cycle followed by 12 cycles at C/4 with 1 h rest before each charge and discharge. This high temperature test was made at 60°C, which is the appropriate condition to investigate possible iron dissolution in non-aqueous electrolytes.

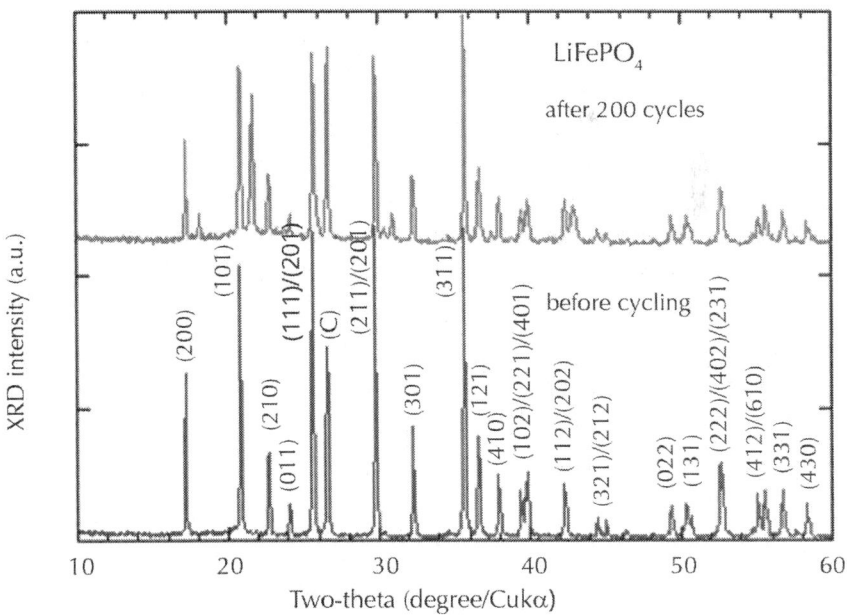

Figure 1: XRD pattern of as-prepared LiFePO$_4$ material (lower curve) and positive electrode after 200 cycles (lower curves). Bragg lines are indexed in the Pmna space group. Notice that the olivine framework remains intact after cycling at 60°C.

The local structure of LFP materials was studied by FTIR in the spectral range 150 - 1400 cm^{-1}.Figure 2 presents the FTIR spectra of several LiFePO$_4$ samples in the low-wavenumber region (300 - 600 cm^{-1}) involving bending modes and in the high-wavenumber domain (600 - 1300 cm^{-1}) involving stretching vibrations. A comparison can be established between impurity-containing samples (A-type) and the optimised phospho-olivine material (B-type). In the low-wavenumber region of the active symmetric and asymmetric (v_2 - v_4) bending modes of P-O bonds, we observe two well-resolved doublets at 349 - 377 and 468 - 500 cm^{-1} for the optimised C-LiFePO$_4$. When an impurity such as Li$_3$PO$_4$ (sample A) is present, the FTIR spectrum displays some modifications in the shape of the doublets and an additional IR band grows at 424 cm^{-1}. The presence of an impurity was also observed in the high-wave number region of the symmetric and asymmetric (v_1 - v_3) modes of PO$_4$ groups. Introduction of LiFeP$_2$O$_7$ is detected by the appearance of two sets of IR bands at 762 and 1180 cm^{-1}. The band at 762 cm^{-1} is due to the symmetric stretching mode of P$_2$O$_7$ pyrophosphate groups, while the high-frequency band at 1180 cm^{-1} is assigned to the vibration of the PO$_3$ terminals [17-19]. These two spectral features are fingerprints of the diphosphate impurity. The inclusion of Li$_3$PO$_4$ is observed by additional spectral features such as enhancement of the asymmetric stretching (v_3) vibration of (PO$_4$)$^{3-}$ oxo-anions. It is worth noting that FTIR measurements (like X-ray diffractometry) are not sensitive to detect of low concentrations of impurities (at the ppm level). The FTIR features show that the carbon does not penetrate significantly inside the LiFePO$_4$ particles, although it is very efficient in reducing Fe^{3+}, which prevents γ-Fe$_2$O$_3$ clustering, thus pointing to a gas-phase reduction process.

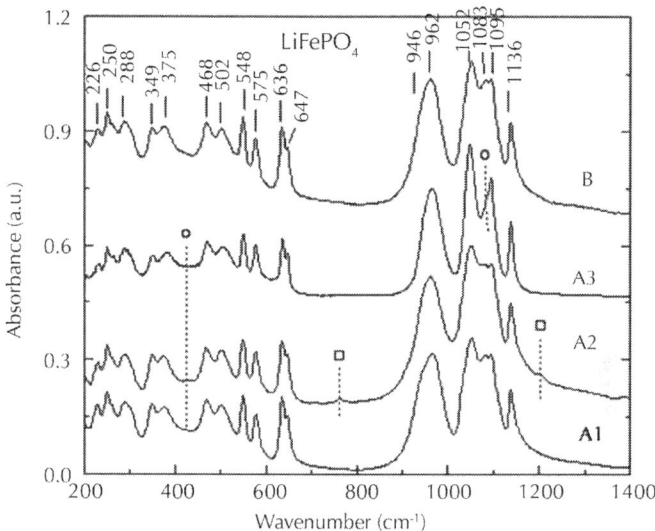

Figure 2: FTIR spectra of LiFePO$_4$ samples. Patterns show bands due to Li$_3$PO$_4$ (o) and LiFeP$_2$O$_7$ (□) impurity. The best LiFePO$_4$ material is the B-10 sample (upper curve).

Figure 3(a) presents the typical SEM image of the prepared LiFePO$_4$ material. The powders are composed of agglomerated crystallites of size 150 nm in average. The existence of the carbonaceous film is also indicated by the bright-field TEM image presented in Figure 3(b). This image displays representative primary particles with a network of carbon in the interstitial grain-boundary region. In the micrograph, the LFP crystallites appear as the darker regions while the carbon coating is surrounding the primary particle as the grayish region.

Iron (III) Nanoclusters in LiFePO$_4$

LiFePO$_4$ undergoes a transition to antiferromagnetic (AFM) order at a Néel temperature T$_N$ = 52 K [20]. Figure 4 shows the temperature dependence of the reciprocal magnetic susceptibility for three samples. While nano-sized ferromagnetic particles were evidenced in previously prepared LiFePO$_4$, such clusters do not exist in the optimized LiFePO$_4$ [10, 11]. Magnetic measurements illustrate that the magnetization M (H) is the superposition of two contributions M (H) = x$_m$H + M$_{extrin}$. The intrinsic part, x$_m$H, is linear in the applied magnetics H and an extrinsic component, M$_{extrin}$ = Nnμ£ (ξ), is easily saturated by the application of H due to ferromagnetic impurities. Here, £ (ξ) is the Langevin function, N is the number of magnetic clusters made of n magnetic moments μ.

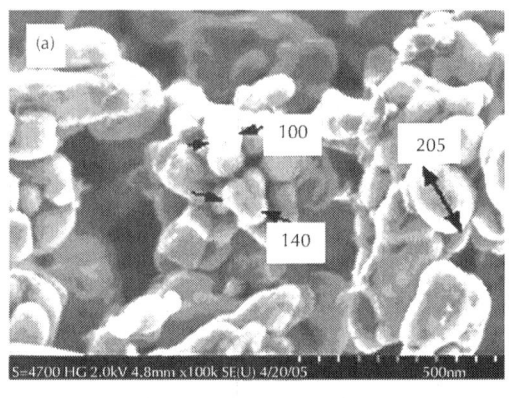

(a)

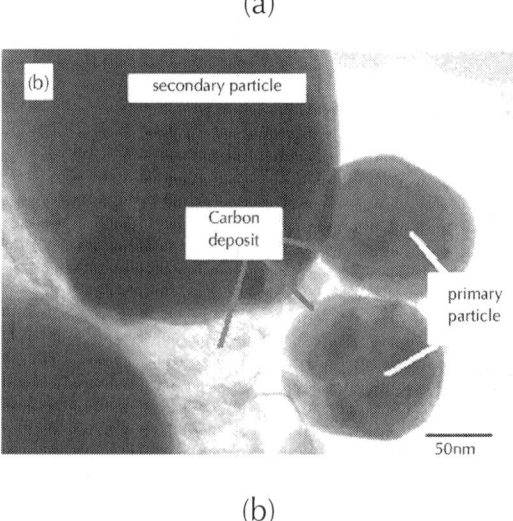

(b)

Figure 3: SEM image (a) of the carbon-coated sample showing the shape of the secondary particles. HRTEM image (b) showing the amorphous carbon layer deposited onto the LiFePO$_4$ crystallite. Particle size is expressed in nm.

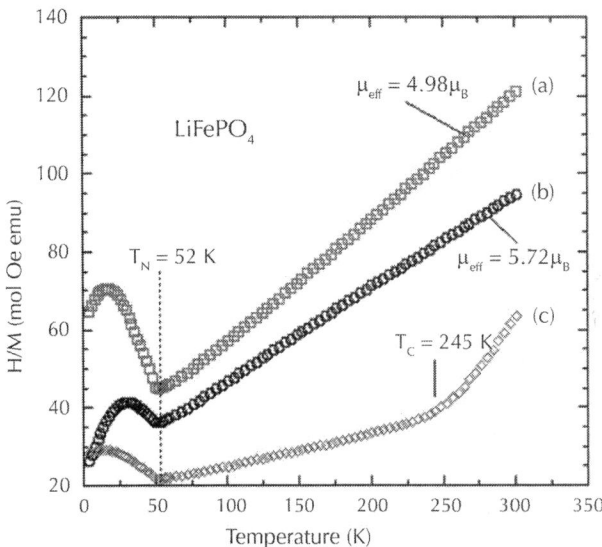

Figure 4: Temperature dependence of the reciprocal magnetic susceptibility of different LiFePO$_4$ samples. (a) Optimized pure LiFePO$_4$; (b) Fe$_2$O$_3$-containing sample; and (c) x^{-1}(T) Fe$_2$P-containing sample.

Analysis of the magnetic properties gives an average separation of the magnetic clusters that is too large for interaction between particles (superparamagnetic model). This hypothesis must be released where the number n of magnetic clusters is so large that magnetic interactions between the ferrimagnetic particles become important [9]. At high fields, M$_{extrin}$ saturates to Nnμ so that this quantity is readily determined as the ordinate at H = 0 of the intersection of the tangent to the magnetization curves at large fields. As a result, we find that Nnμ does not depend significantly on temperature below 300 K. We are in the situation where the cluster magnetization is temperature independent, which amounts to say that the Curie temperature T$_C$ inside the clusters is much larger than 300 K. This is important information on the nature of the ferromagnetic clusters. In particular, this feature precludes the existence of Fe$_2$P clusters in some LFP samples prepared according to a different procedure [10], since the Curie temperature of these clusters is only 220 K. The nature of the strongly ferromagnetic clusters in the present case is most likely maghemite (γ-Fe_2O_3).

It is remarkable from Figure 4 that the A-type sample displays different magnetic features i.e. with a magnetic moment $\mu_{eff} = 5.72\mu_B$, due to the existence of Fe (III) containing impurities. The first consequence is an ambiguity in what is called the magnetic susceptibility x_m since M/H is distinct from dM/dH. The magnetic susceptibility measured with a SQUID at H = 10 kOe shows the non-linearity of the magnetic moments attributed to the presence of γ-SFe_2O_3. The best material shows the lowest Curie constant 3.09 emu·K/mol. The effective magnetic moment $\mu_{eff} = 4.98\mu_B$ is close to theoretical value $4.90\mu_B$ calculated from the spin-only value of Fe^{2+} in its highspin configuration. Departure from the spin-only value may reflect the presence of Fe^{3+} ions and/or an orbitalmomentum contribution from the Fe^{2+} ions [12].

Influence of the Fe_2P Nanoclusters

The electrochemical properties of LFP are known to be sensitive to the mode of preparation and the structural properties [21]. This can be an advantage for potential applications since it allows for an optimization of the material if we can correlate the mode of preparation with the structural and the physical properties. To address this issue, we investigated this relationship in $LiFePO_4$ sample that were grown at different conditions. Undesirable impurities in the lattice can be introduced during the growth process. For instance, the presence of Fe_2P can increase the electronic conductivity, but on the other hand it also decreases the ionic conductivity so that both the capacity and cycling rates are degraded with respect to C-LFP. The presence of a small concentration (>0.5%) of Fe_2P is evidenced in Figure 4 by the appearance of an abnormal x (T) behaviour with the occurrence of a shoulder near $T_C = 265$ K, the Curie temperature of the ferromagnet Fe_2P. Figure 5 shows the Arrhenius plot of the electronic conductivity, σ_{elec}, of three $LiFePO_4$ samples: a pure material, a Fe_2P-containing sample, and a C-LFP. It is obvious that addition of either iron phosphide or carbon enhances greatly σ_{elec} but to the detriment of the capacity for the former compound, as it will be discussed next.

Furthermore, we know that hydrogen, carbon monoxide, or carbon can reduce Fe_2O_3 through different reduction steps that depend on temperature and other physical parameter such as particle size. Although we anticipate that over 1000°C carbon might reduce Fe^{3+} ions or through the formation of CO gas to prevent the formation of γ-Fe_2O_3, other factors may be involved. We believe that the carbon deposition process, which was organic precursors to make C-coated samples, generates a reductive gas such as hydrogen that is more kinetically active and reduces Fe^{3+} impurities in the 400°C - 700°C temperature range used in our studies. This model is also favoured by the fact that the organic precursor is usually mixed with the LFP material or with the LiFePO$_4$ chemical precursors by solution processes at a molecular-size level. However, as presented in Figure 5, the carbon coating greatly enhances the electronic conductivity of the LFP particles that allows high rate for the charge discharge process.

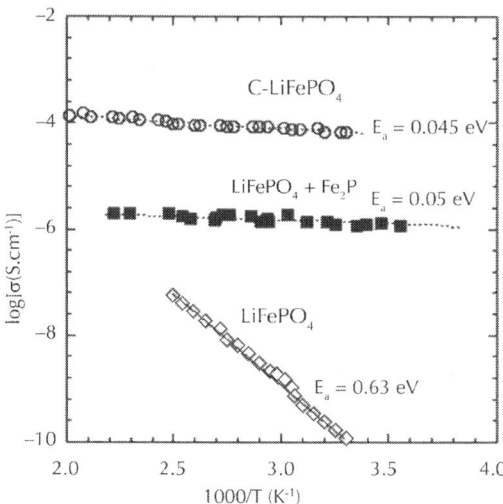

Figure 5: Electronic conductivity of LiFePO$_4$ samples. (a) Pure material; (b) Fe$_2$P-containing sample; and (c) Carboncoated LiFePO$_4$.

Figure 6 displays the electrochemical charge-discharge profiles of Li//LFP cells cycled at room temperature with pure $LiFePO_4$ and with Fe_2P-containing electrode material. It is obvious that at the rate 2C, the capacity retention decreases significantly for the material containing few% of Fe_2P. A close examination was made for the detection of any iron dissolution that could occur after long-term cycling. The analysis of iron species was investigated at the separator/lithium (SL) interface by SEM cross-section (slice view) as shown in Figures 7(a) and (b). The micrograph (Figure 7(a)) obtained from evaluation of the earlier generation material shows the presence of iron islands at the SL interface. Obviously, some iron particles (or ions) migrate through the electrolyte from the $LiFePO_4$ positive electrode to the lithium negative. The net effect of this migration is a large decrease in capacity retention of the Li//LFP cell. Figure 7(b) shows the post-mortem micrograph obtained from tests with an optimised electrode in a Li cell with a lithium foil negative. In this case, there is no iron detected at the SL interface, which remained intact after 100 cycles. In fact, this high performance was possible not only because the optimised synthesis of the LFP powders, but also because of strict control of the structural quality of the materials. Several physical methods were utilized to analyse the local structure and the electronic properties of the phospho-olivine framework.

Electrochemical Properties of Optimized $LiFePO_4$

To study the capacity fade of $LiFePO_4$ at 60°C, we used three different negative electrodes, namely lithium metalgraphite, and $Li_4Ti_5O_{12}$. A lithium metal anode gave exact capacity during charge-discharge process with Li metal excess 2.5 times to $LiFePO_4$ cathode material. Due to the large excess of lithium metal, it was difficult to observe the capacity fade with this anode at 60°C. Graphite anode was 5% of excess to $LiFePO_4$ cathode. This type of anode can detect easily the dissolution of iron because the passivation layer of graphite anode is ionic conductor and electronically insulator; so the dissolution of iron from the cathode side to the anode increases the electronic conductivity of passivation layer of graphite that results on the capacity fade of the cell. LTO has been used because is passivation layer free and zero strain material. The anode has 0% excess to the cathode that gives the high stability of the cycling and also prevents side reactions or reduction of electrolyte to the potential of LTO (1.5 V vs Li^+/Li^0).

Figures 8(a)-(c) present the electrochemical performance of cell in several configurations using LFP as positive electrode. The typical electrochemical profile of the C-LFP/1 mol·L^{-1} $LiPF_6$-EC-DEC/Li cell at 60°C is shown in Figure 8(a). The charge-discharge curves were obtained by cycling at C/4 rate (about 35 mA/g) in the voltage range 2.2 - 4.0 V vs Li^0/Li^+. The optimized LFP exhibits a reversible capacity that is maintained over many charge-discharge cycles. The 10th and 120th cycle shows a similar specific capacity of 160 mAh·g^{-1}. These results illustrate the excellent electrochemical performance of the carbon-coated olivine material. The electrode can be fully charged up to 4 V, which is its most reactive state. This remarkable performance is attributed to the optimized carbon-coated particles and their structural integrity under a large current in the electrode. Even at such a high cycling rate, C-$LiFePO_4$ exhibits rapid kinetics of lithium extraction, and realizes most of its theoretical capacity (170 mAh·g^{-1}). The discharge profile appears with the typical voltage plateau (at ca. 3.45 V vs Li^0/Li^+) attributed to the two-phase reaction of the $(1 - x) FePO_4 + xLiFePO_4$ system.

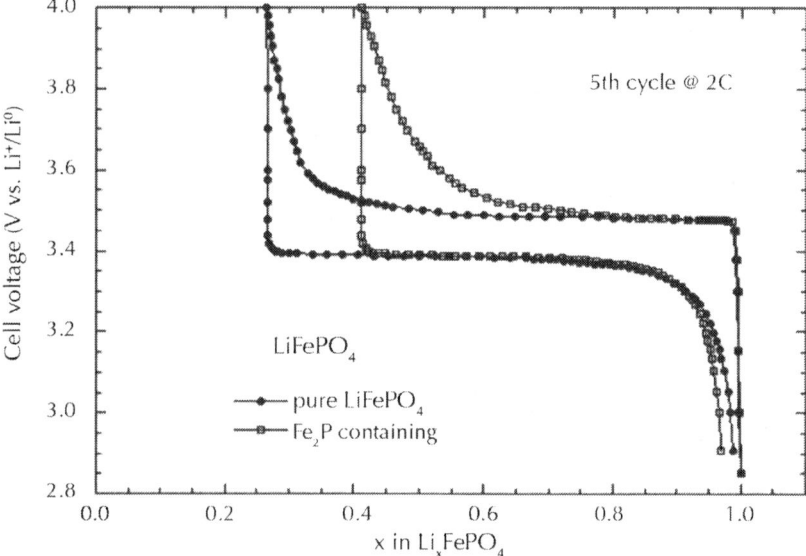

Figure 6: Electrochemical charge-discharge profiles of Li//LiFePO$_4$ cells cycled at room temperature. (a) With pure LiFePO$_4$; and (b) With Fe$_2$P-containing electrode material.

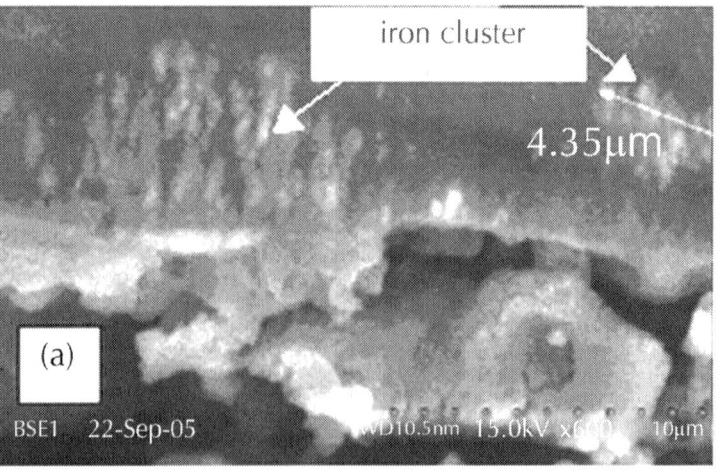

(a)

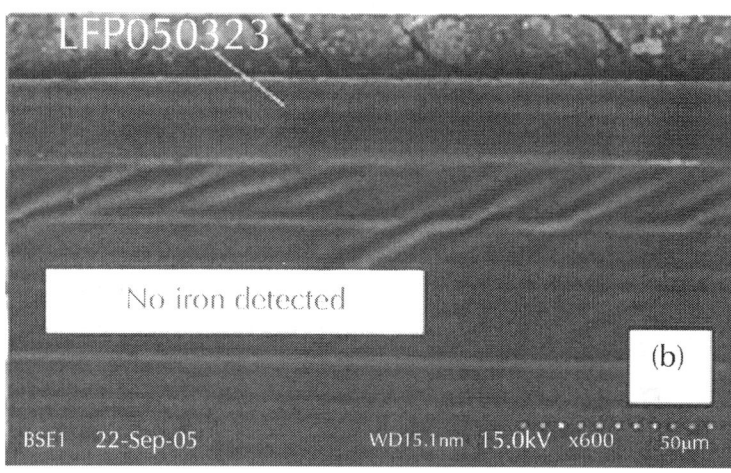

(b)

Figure 7: Post-mortem SEM images of the detection of iron species at the separator/lithium interface. (a) Formation of iron islands at the interface with an earlier generation of LiFePO$_4$; (b) No iron was detected at the surface of Li foil with the optimized LiFePO$_4$.

To investigate the impact of the crystallization of the surface layer of LiFePO$_4$ particles on the electrochemical properties, we have reported the first charge/discharge of a coin cell in Figure 8(b), in which the active cathode element was non-coated LFP heated at 700°C during 4 h and C-LFP for comparison. The same powder (particles of average thickness 40 nm) was used to avoid any size effect of the particles on the electrochemical particles. The experiments were performed at rate C/12 at room temperature. The results show that the capacity of the cell prepared with LFP heat treated at 700°C is very small despite the crystallization of the surface layer of the LiFePO$_4$particles. However, the capacity only reaches ca. 55% of its theoretical value, while, after carbon coating, the capacity of these particles is close to the theoretical value. This is indeed the evidence that, even in the case of nano-scaled particles, the carbon coating alone is far from sufficient to recover the full capacity of C-LFP, for the reasons we have recalled in the introduction. In our previous exploration of the surface properties of the LFP particle, we have shown by Raman spectroscopy that the deposit is a disordered graphite-type carbon [22]. The small amount of carbon (<2 wt %) can be viewed as a film of irregular thickness, 3 nm thick on average, with gaps. The above experimental condition (ca. 60°C) has a severe impact on the kinetics of the Fe^{2+}/Fe^{3+} redox reaction, but the recent report from Hydro-Québec Research Labs showed that this type of C-LFP electrode can be cycled at 60°C without significant capacity loss for over 200 cycles [7]. Optimized particle size in the range 200 - 300 nm agrees well with the average diameter of grains L that validates the characteristic diffusion time $\tau = L^2/4\varpi^2 D^*$ [23], where D^* is the chemical diffusion coefficient of Li^+ ions in the $LiFePO_4$ matrix (typically 10^{-14} $cm^2 \cdot s^{-1}$) when compared with the experimental discharge rate up to 5C.

The electrochemical performance of optimized LFP and LTO electrode materials has been tested separately in half cell with respect to Li metal anode, using the same electrolyte mentioned above. The voltage vs capacity curves recorded under such conditions at 25°C are reported in Figure 8(c) at low C-rate C/24 to approach thermodynamic equilibrium together with the potentialcapacity curve of the LTO//LFP lithium-ion battery. The voltage window is 2 - 4 V for $LiFePO_4$, 1.2 - 2.5 V for $Li_4Ti_5O_{12}$. Note in this figure (and the following ones), we have kept the conventional rule, i.e. the capacity is in mAh per gram of the active element of the cathode. That is the reason why the maximum capacity for the LFP//Li and LFP//LTO cells are the same. For LFP//Li, the first coulombic efficiency is 100 % and the reversible capacity is 148 $mAh \cdot g^{-1}$. For LTO, the first coulombic efficiency is 98% and the reversible capacity is 157 $mAh \cdot g^{-1}$. The well-known plateaus at 3.4 and 1.55 V are characteristics of the topotactic insertion/deinsertion of lithium in the two-phase systems $LiFePO_4$-$FePO_4$ and $Li_4Ti_5O_{12}$- $Li_7Ti_5O_{12}$, respectively.

The electrochemical performance of C-$LiFePO_4$ was tested in various conditions of temperature. At 2C rate, the capacity retention was 153, 136 and 93 $mAh \cdot g^{-1}$ for cells discharged at 60°C, 25°C and −10°C, respectively [24]. The Ragone plots of cell cycled at 25°C and 60°C are shown in Figure 9. The cells were cycled in the potential range 2.5 - 4.0 V. The discharge capacity and electrochemical utilisation, i.e. the ratio discharge/charge, vs cycle number are excellent for the C-LiFePO$_4$/LiPF$_6$- EC-DEC/Li cells. At 10C rate, these Li-ion cells provide coulombic efficiencies 85% at 60°C.

CONCLUSIONS

In this work, we compared the physico-chemical and electrochemical properties of a series of carbon-coated LiFePO$_4$ samples. A major effect of the carbon deposition process has been to reduce Fe^{3+}, most probably through a gas-phase reduction process involving hydro-gen from the organic carbon precursor. The hydrogen prevents formation of γ-Fe$_2$O$_3$ nanoparticles in which iron is in the trivalent state. The magnetic measurements indicated the presence of nanoclusters, i.e. Fe$_2$P or/and γ-Fe$_2$O$_3$ in non-optimized samples. Thus, this study demonstrates that magnetic measurements (combination of M (H) and x (T) data) are beneficial for detecting ferric and/or ferrous impurities, as well as for the quality control of LiFePO$_4$. Electrochemical tests have been conducted under various conditions to assess the influence of the electrolyte on stability and the influence of electrode processing. Post-mortem analysis showed that no iron species were detected at the separator-negative electrode interface in cells with lithium metal, graphite and CLi$_4$Ti$_5$O$_{12}$. This result is attributed to the high quality of the "optimised" LiFePO$_4$, impurity-free materials used as positive electrodes. The discharge capacity and electrochemical utilisation, i.e. the ratio discharge/charge, vs cycle number are excellent for the C-LiFePO$_4$/LiPF$_6$-ECDEC/Li cells. At 10C rate, these Li-ion cells provide coulombic efficiencies 85% at 60°C.

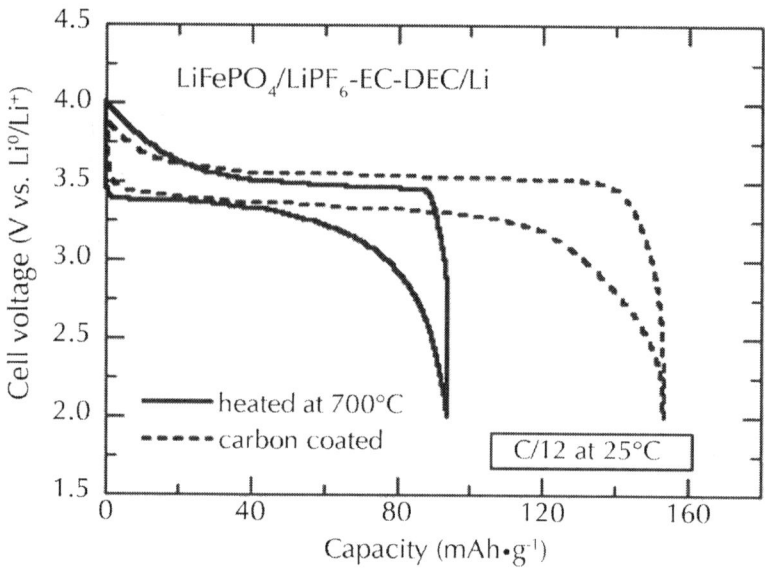

(a)

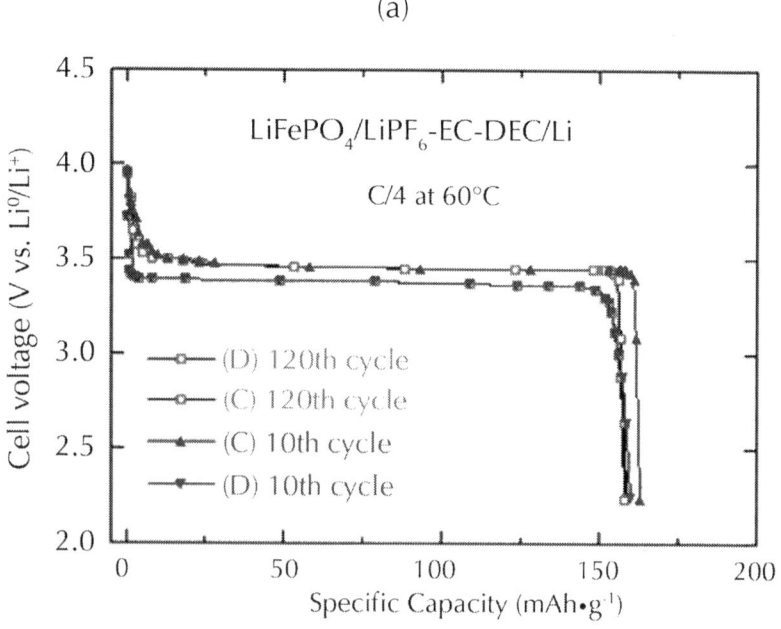

(b)

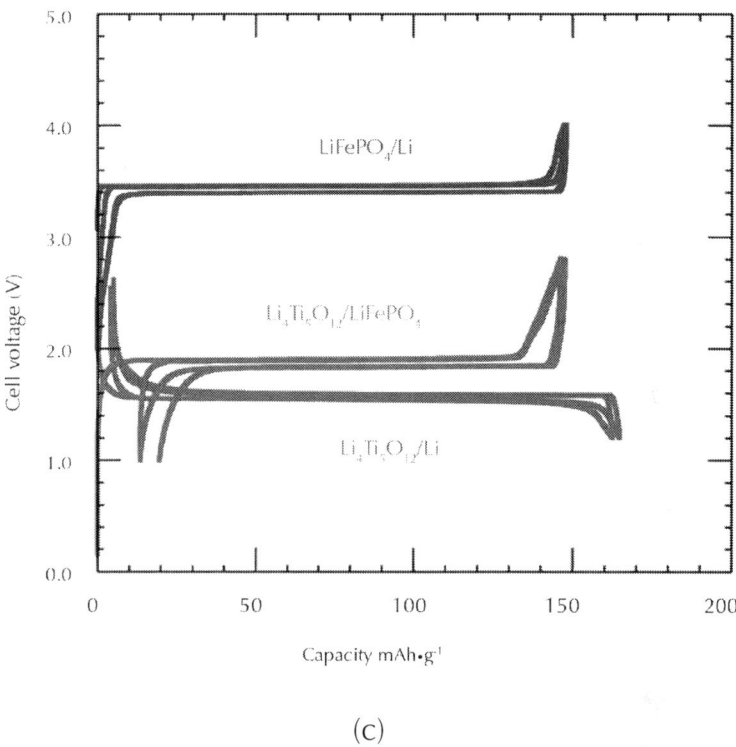

(c)

Figure 8: (a) Electrochemical performance of the C-LiFePO$_4$//Li cell operating at 60°C. Charge-discharge cycling was conducted at the C/4 rate; (b) Electrochemical profiles at C/12 of LiFePO$_4$//Li cells. Positive electrodes were (i) non-coated and heated at 700°C (c) carbon-coated; (c) Voltage-capacity cycle for LiFePO$_4$//Li, Li$_4$Ti$_5$O$_{12}$//Li and Li-ion cell LiFePO$_4$//Li$_4$Ti$_5$O$_{12}$ at C/24 rate. The capacity is in mAh per gram of the positive electrode element (LiFePO$_4$, Li$_4$Ti$_5$O$_{12}$ and LiFePO$_4$, respectively). The larger hysteresis in the LiFePO$_4$//Li$_4$Ti$_5$O$_{12}$ cell comes from the fact that the cell in that case was a button cell instead of the more elaborate 18650-cell, but the plateau at 1.9 V is well observed. All the cells used 1 mol·L^{-1} LiPF$_6$ in EC:DEC (1:1) as electrolyte.

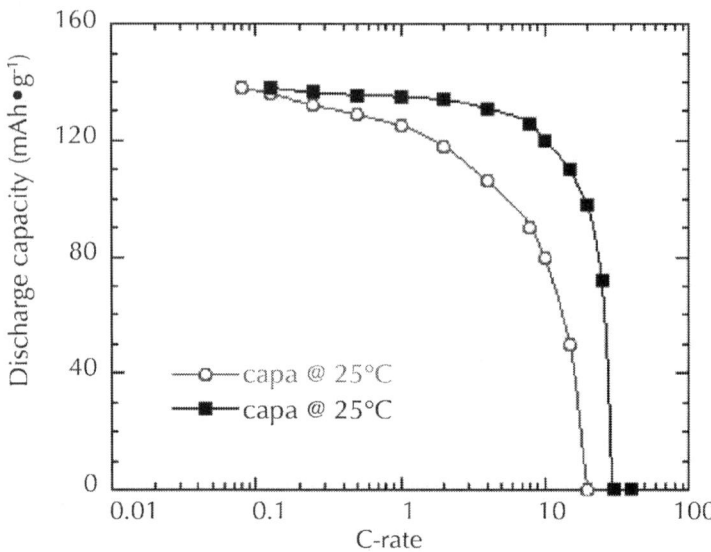

Figure 9: Ragone plots of the C-LiFePO$_4$/LiPF$_6$-EC-DEC/Li cells as a function of the working temperature 25°C and 60°C.

REFERENCES

1.   A. K. Padhi, K. S. Nanjundaswamy and J. B. Goodenough, "Phospho-Olivines as Positive-Electrode Materials for Rechargeable Lithium Batteries," Journal of the Electrochemical Society, Vol. 144, No. 4, 1997, pp. 1188-1194. doi:10.1149/1.1837571

2.   N. Ravet, Y. Chouinard, J. F. Magnan, S. Besner, M. Gauthier and M. Armand, "Electroactivity of Natural and Synthetic Triphylite," Journal of Power Sources, Vol. 97- 98, 2001, pp. 503-507. doi:10.1016/S0378-7753(01)00727-3

3.   N. Ravet, A. Abouimrane and M. Armand, "From Our Readers: On the Electronic Conductivity of PhosphoOlivines as Lithium Storage Electrodes," Nature Materials, Vol. 2, 2003, pp. 702-703. doi:10.1038/nmat1009b

4. N. Ravet, S. Besner, M. Simoneau, A. Vallée, M. Armand and J. F. Magnan, "Electrode Materials with High Surface Conductivity," US Patent No. 6962666, 2005.

5. Y. Hu, M. M. Doeff, R. Kostecki and R. Finones, "Electrochemical Performance of Sol-Gel Synthesized $LiFePO_4$ in Lithium Batteries," Journal of the Electrochemical Society, Vol. 151, No. 8, 2004, pp. A1279-A1285. doi:10.1149/1.1768546

6. S. L. Bewlay, K. Konstantinov, G. X. Wang, S. X. Dou and H. K. Liu, "Conductivity Improvements to SprayProduced $LiFePO_4$ by Addition of a Carbon Source," Materials Letters, Vol. 58, No. 11, 2004, pp. 1788-1791. doi:10.1016/j.matlet.2003.11.008

7. K. Zaghib, V. Battaglia, P. Charest, V. Srinivasan, A. Guerfi and R. Kostecki, Extended Abstract of the International Battery Association & Hawaii Battery Conference, Wailoloa, 9-12 January 2006.

8. A. A. Salah, P. Jozwiak, J. Garbarczyk, F. Gendron, A. Mauger and C. M. Julien, "FTIR Features of Lithium Iron Phosphates Used as Positive Electrodes in Rechargeable Lithium Batteries," 207th ESC Meeting, Québec City, 15- 20 May 2005.

9. M. M. Doeff, Y. Hu, F. McLarnon and R. Kostecki, "Effect of Surface Carbon Structure on the Electrochemical Performance of $LiFePO_4$," Electrochemical and SolidState Letters, Vol. 6, No. 10, 2003, pp. A207-A209. doi:10.1149/1.1601372

10. A. Ait-Salah, A. Mauger, F. Gendron and C. M. Julien, "Magnetic Studies of the Carbothermal Effect on $LiFePO_4$," Physica Status Solidi (a), Vol. 203, No. 1, 2006, pp. R1- R3.

11. A. Ait-Salah, A. Mauger, C. M. Julien and F. Gendron, "Nano-Sized Impurity Phases in Relation to the Mode of Preparation of $LiFePO_4$," Materials Science and Engineering: B, Vol. 129, No. 1-3, 2006, pp. 232-244. doi:10.1016/j.mseb.2006.01.022

12. A. Ait-Salah, A. Mauger, K. Zaghib, J. B. Goodenough, N. Ravet, M. Gauthier, F. Gendron and C. M. Julien, "Reduction Fe^{3+} of Impurities in $LiFePO_4$ from Pyrolysis of Organic Precursor Used for Carbon Deposition," Journal

of the Electrochemical Society, Vol. 153, No. 9, 2006, pp. A1692-A1701. doi:10.1149/1.2213527

13. A. Yamada, S. C. Chung and K. Hinokuma, "Optimized LiFePO$_4$ for Lithium Battery Cathodes," Journal of the Electrochemical Society, Vol. 148, No. 3, 2001, pp. A224-A229.doi:10.1149/1.1348257

14. K. Zaghib, J. Shim, A. Guerfi, P. Charest and K. A. Striebel, "Effect of Carbon Source as Additives in LiFePO$_4$ as Positive Electrode for Lithium-Ion Batteries," Electrochemical and Solid-State Letters, Vol. 8, No. 4, 2005, pp. A207-A210. doi:10.1149/1.1865652

15. K. Zaghib and M. Armand, "Electrode Covered with a Film Obtained from an Aqueous Solution Containing a Water Soluble Inder, Manufacturing Process and Usesthereof," Canadian Patent No. CA 2411695, 2002.

16. K. Striebel, J. Shim, V. Srinivasan and J. Newman, "Comparison of LiFePO$_4$ from Different Sources," Journal of the Electrochemical Society, Vol. 152, No. 4, 2005, pp. A664-A670.doi:10.1149/1.1862477

17. A. Ait-Salah, P. Jozwiak, J. Garbarczyk, K. Benkhouja, K. Zaghib, F. Gendron and C. M. Julien, "Local Structure and Redox Energies of Lithium Phosphates with Olivineand Nasicon-Like Structures," Journal of Power Sources, Vol. 140, No. 2, 2005, pp. 370-375.doi:10.1016/j.jpowsour.2004.08.029

18. R. Bacewicz, P. Woroniecki and J. Garbarczyk, "Raman Scattering in AgI-Ag$_2$O-P$_2$O$_5$Glasses," Physics and Chemistry of Glasses, Vol. 40, No. 3, 1999, pp. 175-176.

19. A. Adamczyk, M. Handke and W. Mozgawa, "FTIR Studies of BPO$_4$·2SiO$_2$, BPO$_4$·SiO$_2$ and 2BPO$_4$·SiO$_2$ Joints in Amorphous and Crystalline Forms," Journal of Molecular Structure, Vol. 511-512, 1999, pp. 141-144. doi:10.1016/S0022-2860(99)00152-0

20. R. P. Santoro and R. E. Newnham, "Antiferromagnetism in LiFePO$_4$," Acta Crystallographica, Vol. 22, 1967, pp. 344-347. doi:10.1107/S0365110X67000672

21. G. Arnold, J. Garche, R. Hemmer, S. Ströbele, C. Vogler and M. Wohlfahrt-Mehrens, "Fine-Particle Lithium Iron Phosphate $LiFePO_4$ Synthesized by a New Low-Cost Aqueous Precipitation Technique," Journal of Power Sources, Vol. 119-121, 2003, pp. 247-251.doi:10.1016/S0378-7753(03)00241-6

22. C. M. Julien, K. Zaghib, A. Mauger, M. Massot, A. AitSalah, M. Selmane and F. Gendron, "Characterization of the Carbon Coating onto $LiFePO_4$ Particles Used in Lithium Batteries," Journal of Applied Physics, Vol. 100, No. 6, 2006, Article ID: 063511.doi:10.1063/1.2337556

23. S. H. Yu, C. K. Park, H. Jang, C. B. Shin and W. Il Cho, "Prediction of Lithium Diffusion Coefficient and Rate Performance by Using the Discharge Curves of $LiFePO_4$ Materials," Bulletin of the Korean Chemical Society, Vol. 32, No. 3, 2011, pp. 852-856.

24. K. Zaghib, P. Charest, M. Dontigny, A. Guerfi, M. Petitclerc and M. Duchesne, "Olivines: 10 Years R & D at Hydro-Québec in Li-Ion Batteries," Rechargeable Lithium and Lithium Ion Batteries Battery/Energy Technology, Washington DC, 7-12 October 2007, Abstract No. 637.

Chapter 6

Application of a Particle Extraction Process at the Interface of Two Liquids in a Drop Column— Consideration of the Process Behavior and Kinetic Approach

Jacqueline V. Erler[1], Tom Leistner[2], and Urs A. Peuker[1]

[1]Institute of Mechanical Process Engineering and Minerals Processing, Technical University Bergakademie Freiberg, Freiberg, Germany
[2]Helmholtz-Institute Freiberg of Resource Technology, Freiberg, Germany

ABSTRACT

The focus of this research is a new type of particle extraction process for the transfer of magnetite nanoparticles from an aqueous to an immiscible organic phase, directly through the liquid-liquid phase boundary in a drop column. The particle extraction process comprises several advantages such as a minimum amount of stabilizing surfactant, no exposure of the particles to a gas atmosphere and with it the avoidance of sintering by capillary forces and a high particle concentration in the receiving phase as well. The study presents experimental results of the characterization of the process environment and the transfer behavior in a drop column. The solution of surfactant in the continuous phase has been investigated during a particle-free phase transfer experiment including the measurements of the total organic carbon (TOC) content and analysis of the size of the stabilized droplets using the laser diffraction spectroscopy. The determination of the transfer fluxes; the mass flows as well as the yield of transferred magnetite by ICP-OES measurements provide information on the impact of interaction of the elementary processes at the phase boundary. Furthermore, the transfer kinetics of the process is described and compared with calculated theoretical values resulting from a kinetic approach.

INTRODUCTION

Because of their special magnetic properties, magnetite nanoparticles have a great potential for many technological applications. Therefore, they are very interesting for a broad range of research areas, for example as magnetic fluids for low friction dynamic gasket systems as well as for the construction of vibration dampers and tweeters [1] [2]. A particular research focus represents the application of magnetite nanoparticles as advanced functional materials for surface coatings and particle composite materials [3] [4].

Possible areas of application for this can be found in the field of reaction engineering [5], in the form of magnetically separable catalyst material as well as in the biomedical sector [6] - [8] . Especially in combination with polymers, stabilized magnetite nanoparticles are required in an organic phase [9] - [12]. However, the nanoparticles are synthesized mainly in an aqueous phase and have to be placed in an organic phase through appropriate procedures. Due to their increased surface area/volume ratio, nanoparticles are susceptible to oxidation and have a great tendency to agglomerate, which may cause a loss of their special magnetic properties. As a consequence, conventional transfer strategies, based on filtration with subsequent drying and redispersion steps, can only be applied conditionally [13]. Therefore, the development of an efficient process to transfer the magnetite nanoparticles from the aqueous to an immiscible organic phase is of great interest.

Emphasis should be put on producing stable colloidal and functionalized particles continuously with a minimum use of surfactants in the liquid organic medium. To demonstrate the continuous phase transfer process via a particle extraction, we design a concept of a miniplant using a drop column as the chosen transfer device. In the literature, the bubble columns with different internals are thoroughly investigated regarding to mass transfer, flow patterns, bubble shapes and hydrodynamics, for example with applications in metallurgical, chemical, bioand petrochemical process industries [14] -[17] . In our study, a column is used in the simplest form without internals. An organic liquid as disperse phase is injected as drops through a distributor into a magnetite nanoparticle suspension, which represents the continuous aqueous phase. Already, in 1968 Lai and Fuerstenau [18] have carried out a liquid-liquid extraction of ultrafine particles to separate a mixture of alumina, water and oil, however they use for their experiments a separatory funnel to put on top of each other the two phases.

In this study, the particle extraction process is investigated and the results are presented with the drop column in a partial recirculation operation. This means that the aqueous phase is stationary and that the organic phase is in recirculation, as shown in Figure 1. This is a first step to the future operation with a closed circuit of both phases.

The process mechanism of the particle transfer in the drop column can be divided into three fundamental steps. At first, the nanoparticles agglomerate partly due to the instable conditions in the aqueous phase as already mentioned in [19]. In the drop column, after injecting the organic phase, no sedimentation takes place due to the stirring effect of the rising liquid drops. Particle drop collision occurs. At the interface, the surfactants from the organic phase interact with the hydrophilic particle surface. The surfactant molecules adsorbe and become chemically grafted [20] [21]. This leads to a hydrophobization and functionalization of the particles, which allows a phase transition of the magnetite. Depending on the surfactant used, stable organic colloids are formed by an excellent deagglomeration of the particles and a physicochemical dispersion due to the strong repulsive potential of the adsorbed surfactant molecules [12].

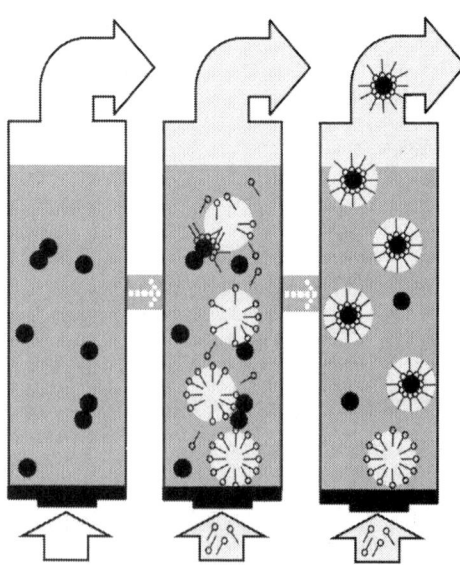

Figure 1: Scheme of the fundamental steps in the liquid-liquid particle extraction process mechanism in the drop column as transfer device.

The aims for the phase transfer process in the drop column are a high efficiency, which means a high yield of transferred magnetite in the organic phase, low transfer times or rather process times and a stable product Furthermore, a stable process in the drop column is necessary. It means a stable drop formation and upward movement with a sufficient coalescence rate, whereby a clearing off of the column is essential for the setup chosen.

MATERIALS AND METHODS

The synthesis of the magnetite nanoparticles with a crystallite size of about 15 nm [21] by a wet-chemical coprecipitation reaction is carried out at 70°C under atmospheric conditions as described by Machunsky et al. [13]. Therefore we applied the precursors iron (II) sulphate heptahydrate and iron (III) chloride hexahydrate purchased from Carl Roth Germany as well as the precipitant ammonium hydroxide solution with an ammonia content of 26% from Sigma.

For the phase transfer experiments in the drop column the pH-conditioning of the aqueous phase is necessary; otherwise emulsion formation in the drop column occurs. The aqueous suspension is conditioned under ambient air by repeated washing with distilled water to a pH value of 4 - 5. For the experiments with oleic acid (OA) the original salt concentration of 39.4 g/l and with ricinoleic acid (RA) a quarter of the original salt concentration is utilized. Subsequently, for both surfactants used a process-pH value of 8.10 ± 0.05 is adjusted.

The research of the liquid-liquid particle extraction process in the drop column is effected by the choice of the surfactants, ricinoleic acid and oleic acid respectively. Both are unsaturated fatty acids (FA) at the 9th C-atom with a carbon chain length of 18 C-atoms. RA has an additional hydroxyl-group at the C-atom C12. The surfactants are of technical grade with 90% purity and purchased from Sigma.

The organic phase as the disperse phase is consisting of the solvent iso-octane provided by Carl Roth Germany with 99.5%

purity as well as the mass fraction of surfactant x_{surf} and a specific amount of surfactant per magnetite $X_{S/M}$. The study presents the results with the parameters x_{surf} = 1.4 mass-% and $X_{S/M}$ = 0.2 g/g. All chemicals are used as received.

The magnetite mass concentration is determined with ICP-OES analyses (inductively coupled plasma optical emission spectroscopy) of Fe with the ICP spectrometer iCAP 6300 from Thermo Fischer Scientific. For the measurements, 5 emission lines of Fe with different wave length ($Fe_{238.2\ nm}$, $Fe_{240.4\ nm}$, $Fe_{259.9\ nm}$, $Fe_{274.6\ nm}$, $Fe_{274.9\ nm}$) are used, which cover the whole mass concentration range. For this, the samples were chemically digested with concentrated hydrochloric acid delivered from Carl Roth Germany. For each emission line a triple determination was performed, resulting with a relative standard deviation of <1%.

The total organic carbon (TOC) content measured with the device from analytikjena multi N/C 2100s is determined by using the difference method. This means the total carbon (TC) content is analyzed. After degassing the inorganic carbon (IC) content is measured and subsequently the TOC content can be calculated. If the IC content is negligible the TC-method is used, in which the TC content is measured directly, and this value corresponds with the TOC content.

For the analysis of the size distribution of stabilized droplets by laser diffraction spectroscopy the spectrometer HELOS, manufactured by Sympatec, is used. The measurement range for this device is between 0.1 - 875 μm.

Experimental Setup

As mentioned the drop column as transfer device is used in a partial recirculation operation, that means the aqueous phase is stationary and the organic phase is in recirculation, as seen in Figure 2. The drop column has a length of 700 mm with an internal diameter of 25 mm. At the bottom of the column the distributor (sparger) as dispersing system is a single metal capillary with an inside diameter of 3.2 mm centrally mounted in a perforated plate. The mixing of the organic phase in the receiver tank is ensured by a mechanical agitator.

Limited by the dimensions of the transfer device a volume of 300 ml of the magnetite suspension, which corresponds to a mass of 6 g magnetite nanoparticles, is filled into the column as continuous phase. The organic phase, which acts as disperse phase is pumped from the receiver tank through the metal capillary into the drop column by a supply system consisting of a peristaltic pump (Ismatec Reglo Analog) and a Tygon-tube with an inside diameter of 3.2 mm. Directly above the opening of the metal capillary, whereby dead zones can be formed aside, the organic phase immediately disintegrates in differently sized drops. These rise through the column in the aqueous phase, due to their lower density. The drops transport the extracted magnetite and thus the nanoparticles are concentrated within the coalesced organic phase. The presence of the solvent drops within the column leads to an expansion of the liquid level, the so-called hold-up.

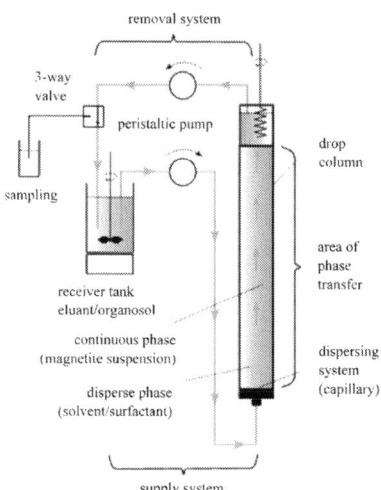

Figure 2: Experimental setup of the miniplant in a partial recirculation operation: aqueous phase stationary and organic phase in recirculation.

The organic phase containing the extracted magnetite particles is collected at the top of the liquid and pumped back to the receiver tank through the removal system. Due to the fact that surfactants are used, the rising drops can form a so called dispersion ribbon at the surface of the continuous phase. This is a layer consisting of droplets and only partial hydrophobized particles, where the coalescence is hindered [22]. The removal system comprises a peristaltic pump and Tygon-tubes with the same properties as in the supply system as well as an automated sampling system with a pneumatic 3-way valve, manufactured by Swagelok, which is controllable via measurement software.

Due to the hydrodynamic dead time the first measuring point can be taken after 85 seconds, because then organosol is existent in the removal system. The average resistence time in the column is about 51 seconds and the volume flow rate is 29 ml/min. The sample volume amount is 2.9 ml and is added at the same time manually as pure organic phase in the receiver tank, hereby keeping the volume of the disperse phase constant during the recirculation. However, after each removed sample volume the amount and mass concentration of magnetite nanoparticles in the system is reduced, which is included in the calculation of $\beta_{magn,rt}(t)$ based on the validation of the system.

For the evaluation of the corrected mass concentration of transferred magnetite nanoparticles in the receiver tank $\beta_{magn,rt}(t)$ a verification of the system is applied taking into consideration the input and output to the receiver tank as well as the residence time distribution in Equation (1):

$$\beta_{magn,rt}\left(t\right) = \beta_{magn,rt,0} \times e^{-\frac{\dot{V}_{disp}}{V_{rt}}t} + \int_{t'=0}^{t'=t} \frac{\dot{V}_{disp}}{V_{rt}}\beta_{magn,in}\left(t\right) \times e^{-\frac{\dot{V}_{disp}}{V_{rt}}(t-t')} dt$$

$$(1)$$

where $\beta_{magn,rt,0}$ is the initial mass concentration of the system, V_{rt} is the volume in the receiver tank, $\dot{V}_{disp}$ is the volume flow of the disperse phase and t' is the auxiliary variable, which describes the time of the entry. By linear interpolation and numeric integration, the Equation (1) can be solved. This results in the following equation:

$$\beta_{magn,rt}(t) = \left(\beta_{magn,rt,0} + \frac{a_i \times V_{rt}}{\dot{V}_{disp}} - b_i \right) \times e^{-\frac{V_{rt}}{\dot{V}_{disp}}t} + a_i t + b_i - \frac{a_i \times V_{rt}}{\dot{V}_{disp}}$$

(2)

where a_i as well as b_i are formed by linear interpolation between the removed samples.

To characterize the process environment we have investigated the procedures in the drop column by the method of a particle-free operation, whereby the changes in the column can be observed. At defined time points the pumps were switched off and the sampling is effected from the continuous phase. In dependence from surfactants used the different mechanisms of drop formation of the disperse phase in the column is demonstrated in Figure 3. With ricinoleic acid as surfactant (Figure 3(a)) the disperse phase disintegrates in a multitude of very small droplets and turbidity appears in the column at once, which spreads in the whole column within a few minutes. In contrast, with oleic acid (Figure 3(b)) as surfactant a single periodic drop formation [23] occurs resulting in larger and significantly less drops without turbidity.

RESULTS AND DISCUSSION

As mentioned above the both surfactants used have a different process behavior with varying consequences to the particle extraction process. The turbidity in the column with ricinoleic acid as surfactant does not influence the process stability of the phase transfer, because the disperse phase can be performed in recirculation without hindrance. The reason for the formation of small emulsion droplets is due to the mass transport of ricinoleic acid through the phase boundary. This is possible, because ricinoleic acid is soluble in the organic as well as in the salts and ammonia containing aqueous phase owing to the additional hydroxyl group. This effect of the spontaneous formation of an emulsion has been described relating to a pH value dependency of Stackelberg, 1949 [24] . In our case an imbalance prevails in the column at the interface of the rising drops, since the ricinoleic acid is dissolved in the organic phase, only at beginning of the process. The system aspires an equalization of concentration by the transport of ricinoleic acid through the phase boundary into the aqueous phase. The resulting convection currents cause deformations and lead to the formation of very small droplets. Figure 4 outlines these stabilized iso-octane droplets with a median value of 5 - 10 µm measured by laser diffraction spectroscopy. They can only be formed if ammonium ricinoleates (carboxylates) exist at the interface by dissolving of ricinoleic acid in the aqueous phase, which definitely reduces the interfacial tension [25] . Furthermore, the smallest median value of the stabilized iso-octane droplets is reached after 20 minutes of defined time points. Due to the low buoyant force of these droplets, the turbidity is appearing in the column.

To prove the solution of surfactants from disperse into continuous phase the TOC content in the aqueous phase is determined and plotted as a function of the time, as shown in Figure 5. By additional centrifugation of the aqueous phase, a potential carbon signal of emulsified solvent can be excluded (light gray columns as short-time trial and dark gray columns as long-time trial). By using ricinoleic acid (Figure 5, top) as surfactant, a significant increase of the TOC content can be observed. After 5 minutes process time the aqueous phase exhibits an amount of dissolved ricinoleic acid in the range of 1.6 - 2.4 g/l independent of the time limit of experimental procedure. Whereas oleic acid (Figure 5, bottom) is used, the poor ability of dissolving within the aqueous phase is expressed considerably.

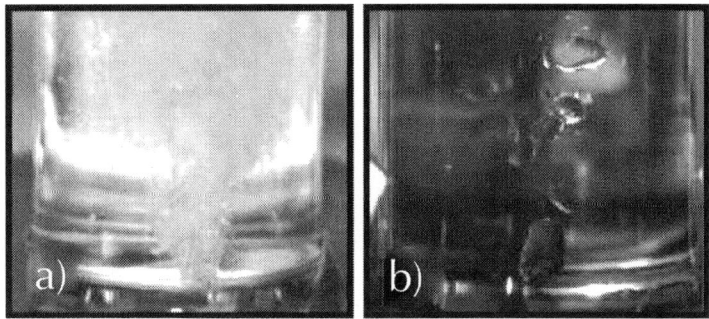

Figure 3: Drop formation of the disperse phase in the column directly above the distributor with ricinoleic acid (a) and oleic acid (b) as surfactant at beginning of the process.

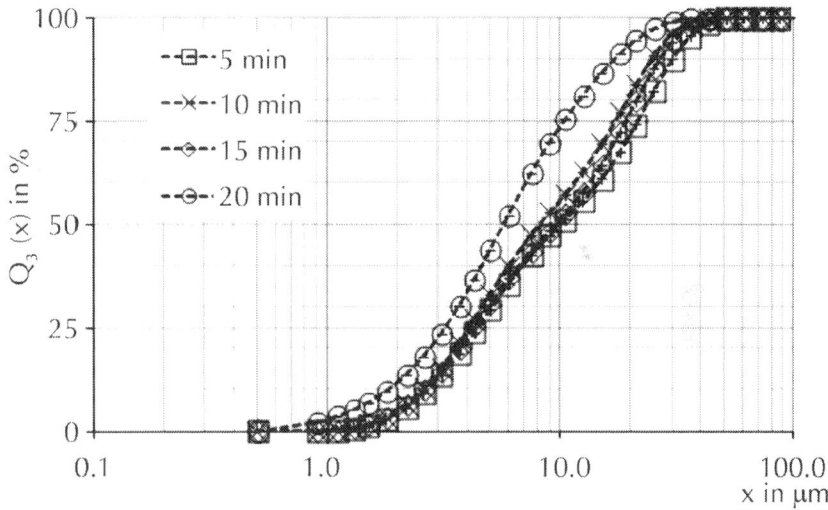

Figure 4: Size distribution of the stabilized iso-octane droplets in the aqueous phase with ricinoleic acid as surfactant using laser diffraction spectroscopy.

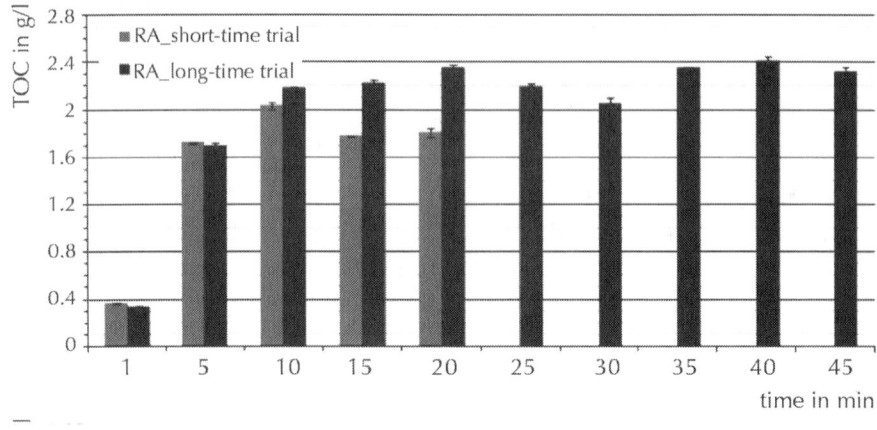

(a)

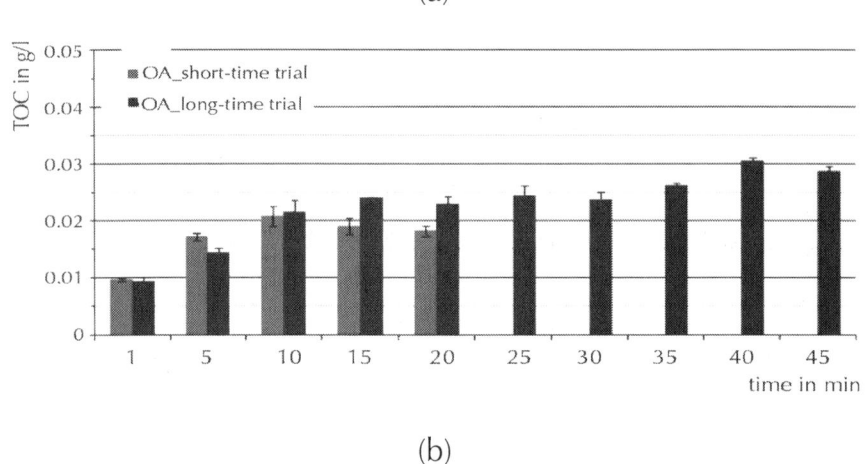

(b)

Figure 5: Total organic carbon (TOC) content at defined time points of sampling in the aqueous phase by centrifugation of the partly emulsified aqueous phase (light gray columns as short-time trials and dark gray columns as longtime trials) with ricinoleic acid (top) and oleic acid (bottom) as a surfactant.

So the TOC values of oleic acid actually dissolved in the aqueous phase are nearly in the range of 0.01-0.03 g/l. Comparing the average values of the mentioned ranges (2 g/l for ricinoleic acid and 0.02 g/l for oleic acid), it can be recognized that in contrast to oleic acid approximately the 100-fold amount of ricinoleic acid is dissolved in the aqueous phase irrespective of time. Furthermore, experiments are performed with a pH-indicator bromothymol blue in the continuous phase, as demonstrated in Figure 6(a), which has a transition point at pH 7.6 (color change from blue to green). For either surfactant an initial process-pH value of 8.1 is adjusted, similar to the conditioned aqueous phase for the phase transfer of magnetite. Only with ricinoleic acid in the disperse phase a change from blue to green is observed in the continuous phase (Figure 6(b)). This indicates that the pH value has decreased below 7.6. The reduction in the concentration of ammonium ions is initiated by the formation of ammonium ricinoleates at the liquid-liquid interface due to the dissolving of ricinoleic acid in the aqueous phase. After the recirculation of the disperse phase with oleic acid as surfactant the color of the continuous phase has changed to light blue (Figure 6(c)), because significantly less of oleic acid is dissolved in the aqueous phase. As a result the formation of ammonium oleates at the interface is lower, which only leads to a slight change of pH in the aqueous phase.

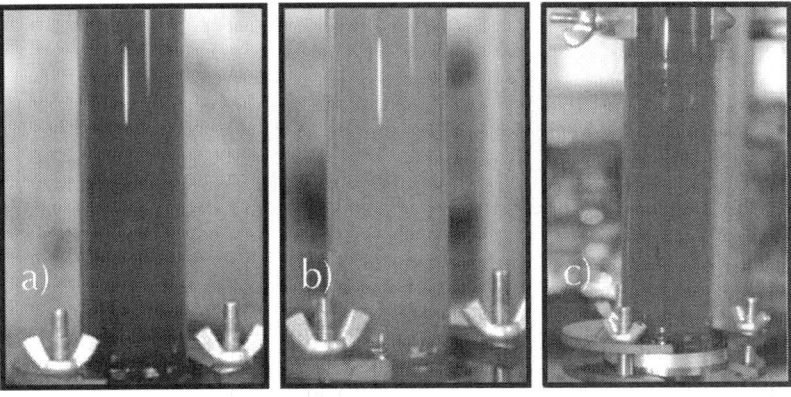

Figure 6: Characterization of the pH value changes in the continuous phase using the indicator bromothymol blue (a) after recirculation of the disperse phase with ricinoleic acid (b) and oleic acid (c) as a surfactant.

This finding implies that more than one process is relevant at the phase boundary.

For the characterization of the transfer behavior the yield of transferred magnetite particles $\phi_{magn,trans}$ is plotted against the process time in Figure 7. This parameter is determined, from the time dependent corrected magnetite mass concentration in the receiver tank $\beta_{magn,rt}(t)$ based on the validation of the system in Equation (1) relating to the theoretical applied magnetite mass concentration for the phase transfer $\beta_{magn,PT,theeor}$ in the following equation:

$$\phi_{magn,trans} = \frac{\beta_{magn,rt}(t)}{\beta_{magn,PT,theor}}$$

(3)

Based on the short-time trial with the process time limit of about 10 minutes it can be observed excellently, that the phase transfer with ricinoleic acid as a surfactant takes place rapidly. The increase of the yield of transferred magnetite is nearly twice the amount compared to oleic acid. This is consistent with the significant increase in TOC content, because by dissolving of ricinoleic acid in the aqueous phase the drops disintegrate more easily and thus an additional phase boundary can be formed for the phase transfer process. Due to the lower solution of oleic acid in the aqueous phase the available phase interface is smaller, which leads to longer process times for the phase transfer of magnetite. This also can be observed during the long-time trials with both surfactants. With ricinoleic acid it can be demonstrated, that already after 15 minutes the yield of transferred magnetite amounts to approximately 90% and it is kept constant about over time, up to 165 minutes. However, with oleic acid the transfer is completed after nearly 270 minutes and the maximum with a yield of transferred magnetite of about 86%. Furthermore, the clearing off of the column with ricinoleic acid as surfactant is obtained after about 6 minutes, whereas with oleic acid it can be achieved only after 160 minutes.

For the characterization of transfer kinetics, the mass flow of transferred magnetite particles $\dot{m}_{magn,trans}$ is plotted against the process time in Figure 8. This parameter is determined, by using the constant volume flow of the disperse phase $\dot{V}_{disp}$ and the difference in mass concentration of transferred magnetite between output and input of the disperse phase $\Delta\beta_{magn,out,in}$ in Equation (4).

$$\dot{m}_{magn,trans} = \dot{V}_{disp} \times \Delta\beta_{magn.out.in} \left(t \right) \tag{4}$$

The plot of the mass flow with ricinoleic acid as a surfactant is conspicuously distinguished by the increase from the first measuring point at 85 seconds till the maximum of the mass flow at a process time between 4 - 5 minutes. During this initial phase the ricinoleic acid concentration in the aqueous phase still rises, which limits the mass flux of magnetite. Subsequently, the mass flow is described with a kinetic of 1st order as it is also identified with oleic acid as surfactant, what needs to be proved.

In the present case the product of functionalized magnetite particles (B) are formed from the precursor of pure magnetite nanoparticles (A) in Equation (5).

$$Fe_3O_{4\,pure.aqueous}\left(A \right) \rightarrow FA@\,Fe_3O_{4organic}\left(B \right) \tag{5}$$

It is supposed that the already transferred magnetite concentration in the organic phase does not influence the transfer process. With this assumption the drop column can be seen as conventional tank reactor [26] , in which the concentration of magnetite is reduced by the transfer reaction.

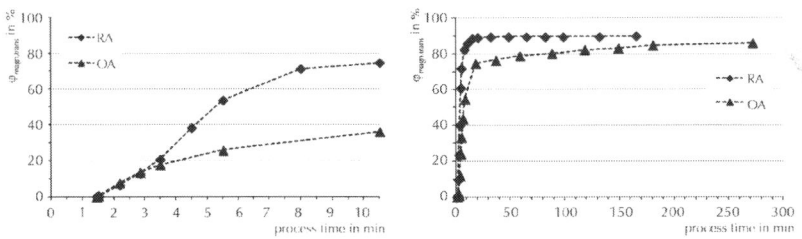

Figure 7: Yield of transferred magnetite in dependency of the process time for ricinoleic acid and oleic acid as a surfactant during short-time and long-time trials.

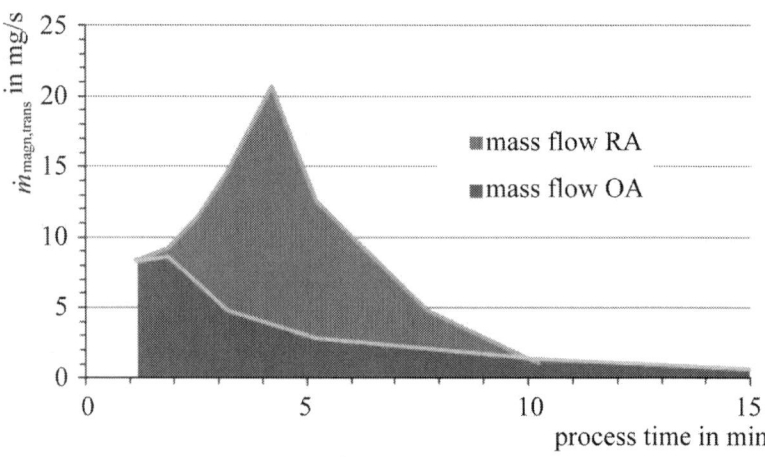

Figure 8: Demonstration of the time dependent mass flow of transferred magnetite particles dependent on the surfactant used.

Thereby, the differential temporary law of the reaction can be established in Equation (6). This indicates, that the temporal change of mass concentration $d\beta_A / dt$ is proportional to the currently existing mass concentration β_A.

$$\frac{d\beta_A}{dt} = -k\beta_A$$

(6)

where k is the rate constant of reaction (1/s) of the 1st order and the product $k\beta_A$ is the rate of consumption of A. Furthermore, the half-life $t_{1/2}$ for the reaction of both surfactants can be determined from the experimental data. The half-life is the time, at which half of the precursor is dissipated. By means of Figure 9 it can be distinguished, that this corresponds to the point of intersection, between the normed mass concentration of the precursor β'_A and the product β'_B. Due to the comparability a dimensionless representation of the mass concentration is chosen by the scaling in Equation (7).

$$\beta' = \frac{\beta_{magn,ICP}}{\beta_{max}}$$

(7)

The relative change in the mass concentration $d\beta_A / \beta_A$ in Equation (8) is proportional to the temporal change dt. It can be calculated from the experimental data by the ICP-OES measurements and is demonstrated in Figure 10 (plots using filled symbols).

$$\frac{d\beta_A}{\beta_A} = -kdt$$

(8)

By integration from t=0 till end time t as well as from initial mass concentration β^0_A till $\beta_A(t)$ and additional by the use of logarithmic, the integral temporary law is achieved in the following equation:

$$\beta_A = \beta_A^0 \times e^{-kt}$$

(9)

resulting in the determination of the theoretical mass concentration β_A and using the Equation (10) for the calculation of k for both surfactants derived from Equation (9)

$$k = \frac{\ln 2}{t_{1/2}}$$

(10)

Hence, the theoretical relative change in the mass concentration can be calculated and compared to the experimental calculated plots (Figure 10).

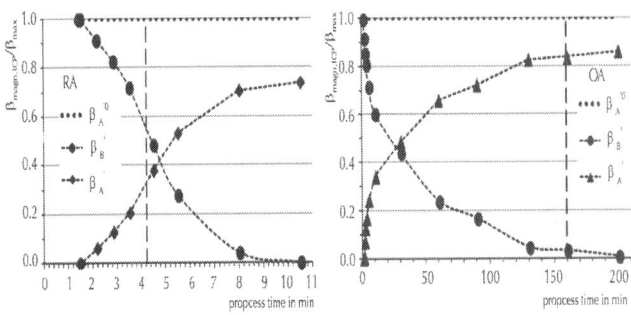

Figure 9: Dependency of the half-life of surfactant used presented by the normed mass concentration plots—with ricinoleic acid (left) the half-life

is 4.78 minutes, whereas the half-life with oleic acid (right) is 27.2 minutes. The dashed line indicates the clearing off of the column.

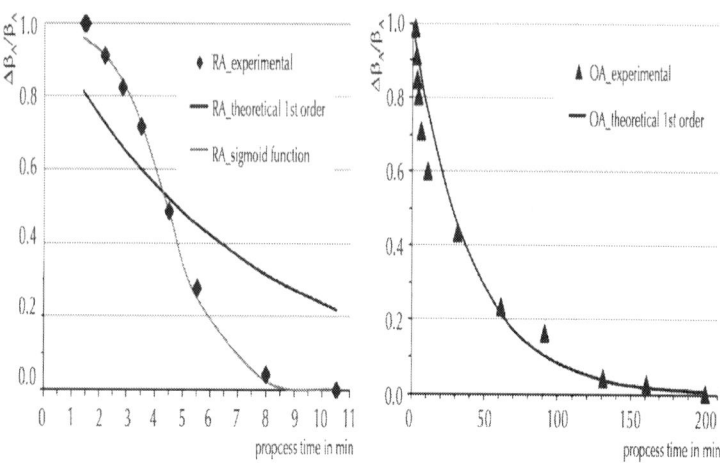

Figure 10: Demonstration of the experimental as well as theoretical calculated relative change in the mass concentration in dependency of the process time with ricinoleic acid (left) and oleic acid (right) as a surfactant.

Based on the comparison in Figure 10 it can be recognized, that the plots between the experimental and theoretical calculated relative change in mass concentration agree only for oleic acid. Therefore, the particle extraction process in partial recirculation operation of the transfer device with oleic acid as a surfactant can be approximately described as a reaction of 1st order. This indicates that the dissolution of oleic acid from the organic in the aqueous phase does not influence the process kinetics. On the contrary, the assumption of a 1st order reaction cannot be used to describe the process response if ricinoleic acid is applied as a surfactant. The course of the experimental relative change in the mass concentration resembles a s-shaped curve. Therefore to characterize the transfer kinetics we can use a sigmoidor a logistic function, respectively, which is applied for approximating saturation processes and also dose-response-systems in Pharmacology [27] . Thus, the particular process behavior for ricinoleic acid can be described using the following equation:

$$\frac{\Delta\beta_A}{\beta_A} = f(t) = A_1 + \frac{A_2 - A_1}{1 + 10^{(M-t)*p}}$$

(11)

Where by A_1 and A_2 the theoretical lower and upper thresholds of the curve. Due to the fact that possible values for $\Delta\beta_A / \beta_A$ lie in the range {0...1}, A_1 and A_2 can be defined with 0 and 1, respectively. Furthermore, Equation (11) has a negative first derivation and, more important, one point of inflection, which demonstrates the local maximum of the mass flow at a process time between 4 - 5 minutes (Figure 8). The point of inflection is represented in Equation (11) by the parameter M, which is estimated through linear interpolation. The parameter p describes the slope of the curve and is numerically calculated by iteration. Thus, the function obtained is illustrated by the grey line in Figure 10, which accurately fits to the experimental data. This demonstrates that the additional mass transport due to dissolving of ricinoleic acid from the organic into the aqueous phase has an important influence on the kinetics of the particle extraction process within the column. The surfactant mass transfer can be assumed as the dose for the process, which predominates in the first minutes of the phase transfer when equilibrium of ricinoleic acid concentration in either phase is not reached. Consequently, the response is represented by the functionalization and subsequently the phase transfer of magnetite nanoparticles. Thus, it can be finally stated out that the particle extraction process is controlled by process imbalance.

CONCLUSIONS

In this work, the particle extraction process with a drop column as transfer device in partial recirculation operation has been investigated. The material parameter of the suspension are chosen to be constant with surfactant concentration x_{surf} = 1.4 mass-% and the optimal specific amount of surfactant per magnetite $X_{S/M}$ = 0.2 g/g. For these parameters, secondary effects like water inclusion and emulsion formation can be excluded, because a stable process behavior with a complete clearing off the column as well as a sufficient coalescence rate can be achieved. Based on the modeling of the system, we are able to evaluate the transfer fluxes. Thereby, yields of transferred magnetite nanoparticles are obtained >80%. Furthermore, for the characterization of the process environment, we have chosen a particle-free phase transfer as an indicator. In combination with optical changes in the column as well as measurements of the total organic carbon (TOC) content and the size distribution using laser diffraction spectroscopy, we have proven that ricinoleic acid helps disintegrating iso-octane. Droplets with a median particle size of 5 - 10 μm are formed. This is particular difference in comparison to oleic acid and this is due to the chemical structure of the surfactants. Therefore, it is possible that additional phase interfaces are formed for the phase transfer process, which further determines the transfer times of particles by reduction of the interface tension. As a consequence of this, we have described the interaction of the processes and procedures at the phase boundary. Thereby, we have a mass transport of the surfactant from disperse into continuous phase on the one hand, which determines the process time for the phase transfer, and simultaneously the phase transfer of the magnetite nanoparticles on the other hand. This interconnection is reflected in the transfer kinetics of either surfactant. Finally, we have determined that the particle extraction process with oleic acid as a surfactant can be estimated as a reaction of 1st order. Thus, the influence of the surfactant mass transfer of the process kinetics is negligible. However, in the case of ricinoleic acid as surfactant, another approximation has to be used to describe the particular process behavior. This is represented by a sigmoid function in terms of a dose response curve. Therefore, the mass transport of the surfactant due to dissolving strongly influences the process kinetics.

Advanced studies in the process development of a continuous liquid-liquid phase transfer to obtain highquality organosols will be presented soon regarding to produce stable colloids and interaction between the surfactants and organic phase used.

ACKNOWLEDGMENTS

The authors would like to thank the German Research Foundation (Deutsche Forschungsgemeinschaft DFG) for financial support by grant PE1160/6-3 and we give thanks to Andre' Rieger as well as the laboratory technicians from Institute of Thermal Process Engineering, Environmental and Natural Products Process Engineering of TU Bergakademie Freiberg for the TOC measurements.

REFERENCES

1. Scherer, M. and Figueiredo Neto, A.M. (2005) Ferrofluids: Properties and Applications. Brazilian Journal of Physics, 35, 718-727. http://dx.doi.org/10.1590/S0103-97332005000400018

2. Hurlebaus, S. and Gaul, L. (2006) Smart Structure Dynamics. Mechanical Systems and Signal Processing, 20, 255-281. http://dx.doi.org/10.1016/j.ymssp.2005.08.025

3. Dallas, P., Georgakilas, V., Niarchos, D., Komninou, P., Kehagias, T. and Petridis, D. (2006) Synthesis, Characterization and Thermal Properties of Polymer/Magnetite Nanocomposites. Nanotechnology, 17, 2046-2053. http://dx.doi.org/10.1088/0957-4484/17/8/043

4. Teja, A.S. and Koh, P.-Y. (2009) Synthesis, Properties and Applications of Magnetic Iron Oxide Nanoparticles. Progress in Crystal Growth and Characterization of Materials, 55, 22-45. http://dx.doi.org/10.1016/j.pcrysgrow.2008.08.003

5. Hickstein, B. and Peuker, U.A. (2009) Modular Process for the Flexible Synthesis of Magnetic Beads—Process and Product Validation. Journal of Applied Polymer Science, 112, 2366-2373. http://dx.doi.org/10.1002/app.29655

6. Banert, T. and Peuker, U.A. (2007) Synthesis of Magnetic Beads for Bio-Separation Using the Solution Method. Chemical Engineering Communications, 194, 707-719. http://dx.doi.org/10.1080/00986440600992750

7.   Laurent, S., Forge, D., Port, M., Roch, A., Robic, C., Vander Elst, L. and Muller, R.N. (2008) Magnetic Iron Oxide Nanoparticles: Synthesis, Stabilization, Vectorization, Physicochemical Characterization and Biological Application. Chemical Reviews, 108, 2064-2110. http://dx.doi.org/10.1021/cr068445e

8.   Mahmoudi, M., Sant, S., Wang, B., Laurent, S. and Sen, T. (2011) Superparamagnetic Iron Oxide Nanoparticles (SPIONs): Development, Surface Modification and Applications in Chemotherapy. Advanced Drug Delivery Reviews, 63, 24-46. http://dx.doi.org/10.1016/j.addr.2010.05.006

9.   Banert, T. and Peuker, U.A. (2006) Preparation of Highly Filled Super-Paramagnetic PMMA-Magnetite Nano Composites Using the Solution Method. Journal of Material Science, 41, 3051-3056. http://dx.doi.org/10.1007/s10853-006-6976-y

10.   Kirchberg, S., Rudolph, M., Ziegmann, G. and Peuker, U.A. (2012) Nanocomposites Based on Technical Polymers and Sterically Functionalized Soft Magnetic Magnetite Nanoparticles: Synthesis, Processing, and Characterization. Journal of Nanomaterials, 2012, Article ID: 670531. http://dx.doi.org/10.1155/2012/670531

11.   Rudolph, M. and Peuker, U.A. (2011) Coagulation and Stabilization of Sterically Functionalized Magnetite Nanoparticles in an Organic Solvent with Different Technical Polymers. Journal of Colloid and Interface Science, 357, 292-299. http://dx.doi.org/10.1016/j.jcis.2011.02.043

12.   Rudolph, M. and Peuker, U.A. (2012) Phase Transfer of Agglomerated Nanoparticles—Deagglomeration by Adsorbing Grafted Molecules and Colloidal Stability in Polymer Solutions. Journal of Nanoparticle Research, 14, 990. http://dx.doi.org/10.1007/s11051-012-0990-6

13.   Machunsky, S. and Peuker, U.A. (2007) Liquid-Liquid Interfacial Transport of Nanoparticles. Physical Separation in Science and Engineering, 2007, Article ID: 34832. http://dx.doi.org/10.1155/2007/34832

14. Youssef, A.A., Al-Dahhan, M.H. and Dudukovic, M.P. (2013) Bubble Columns with Internals: A Review. International Journal of Chemical Reactor Engineering, 11, 1-55.http://dx.doi.org/10.1515/ijcre-2012-0023

15. Shaikh, A. and Al-Dahhan, M. (2007) A Review on Flow Regime Transition in Bubble Columns. International Journal of Chemical Reactor Engineering, 5, 1-68.

16. Vecer, M., Lestinsky, P., Wichterle, K. and Ruzicka, M. (2012) On Bubble Rising in Countercurrent Flow. International Journal of Chemical Reactor Engineering, 10, 1-19.http://dx.doi.org/10.1515/1542-6580.2995

17. Hadavand, L. and Fadavi, A. (2013) Effect of Vibrating Sparger on Mass Transfer, Gas Holdup, and Bubble Size in a Bubble Column Reactor. International Journal of Chemical Reactor Engineering, 11, 1-10. http://dx.doi.org/10.1515/ijcre-2012-0094

18. Lai, R.W.M. and Fuerstenau, D.W. (1968) Liquid-Liquid Extraction of Ultrafine Particles. Transactions of the American Institute of Mining, Metallurgical, and Petroleum Engineers, 241, 549-556.

19. Erler, J., Machunsky, S., Grimm, P., Schmid, H.-J. and Peuker, U.A. (2013) Liquid-Liquid Phase Transfer of Magnetite Nanoparticles—Evaluation Of Surfactants. Powder Technology, 247, 265-269. http://dx.doi.org/10.1016/j.powtec.2012.09.047

20. Zhang, L., He, R. and Gu, H.-C. (2006) Oleic Acid Coating on the Monodisperse Magnetite Nanoparticles. Applied Surface Science, 253, 2611-2617.http://dx.doi.org/10.1016/j.apsusc.2006.05.023

21. Rudolph, M., Erler, J. and Peuker, U.A. (2012) A TGA-FTIR Perspective of Fatty Acid Adsorbed on Magnetite Nanoparticles—Decomposition Steps and Magnetite Reduction. Colloid and Surfaces A: Physicochemical and Engineering Aspects, 397, 16-23.http://dx.doi.org/10.1016/j.colsurfa.2012.01.020

22. Blaß, E. (1988) Bildung und Koaleszenz von Blasen und Tropfen. Chemie Ingenieur Technik, 60, 935-947. http://dx.doi.org/10.1002/cite.330601203

23. Räbiger, N. and Schlüter, M. (2006) Bildung und Bewegung von Tropfen und Blasen. In: VDI Wärmeatlas, Springer, Berlin, 1-15.

24. Stackelberg, M.V. (1949) Spontane Emulgierung Infolge Negative Grenzflächenspannung. Kolloid-Zeitschrift, 115, 53-66. http://dx.doi.org/10.1007/BF01501433

25. Kubatta, E.A. and Rehage, H. (2009) Characterization of Giant Vesicles Formed by Phase Transfer Processes. Colloid and Polymer Science, 287, 1117-1122.http://dx.doi.org/10.1007/s00396-009-2083-3

26. Levenspiel, O. (1999) Chemical Reaction Engineering. John Wiley & Sons, Inc., Hoboken.

27. Chapman, D.G., King, G.G., Berend, N., Diba, C. and Salome, C.M. (2010) Avoiding Deep Inspirations Increases the Maximal Response to Methacholine Without Altering Sensitivity in Non-Asthmatics. Respiratory Physiology and Neurobiology, 173, 157-163.http://dx.doi.org/10.1016/j.resp.2010.07.011

Determination if in Situ Stresses and Elastic Parameters from Hydraulic Fracturing Tests by Geomechanics Modeling and Soft Computing

Shike Zhang[a] and Shunde Yin[b]

[a]School of Civil Engineering and Architecture, Anyang Normal University, Anyang, Henan 45500, China
[b]Department of Chemical & Petroleum Engineering, University of Wyoming, Laramie, WY82071, USA

ABSTRACT

Hydraulic fracturing is the most effective technology for determination of the minimum horizontal in situ stress, $_{h'}$ in a rock formation. The maximum horizontal in situ stress, $_{H'}$ is often determined by the minimum horizontal in situ stress, breakdown pressure, Young's modulus E and Poisson's ratio v with elastic rock behavior assumed. In this paper, a pressure back–analysis method is proposed for determination of these parameters (e.g., $\sigma_{H'}$ $\sigma_{h'}$ E, v) based on borehole pressures monitored in a hydraulic fracturing test. In the proposed method, an artificial neural network (ANN) is used to represent the relationship between maximum and minimum horizontal in situ stresses, elastic parameters and borehole pressure values; a forward model is applied to perform 2-D numerical simulation of a hydraulic fracturing process to create necessary training and testing samples for the ANN model; the genetic algorithm (GA) is employed to search the set of unknown in situ stresses and elastic parameters in a global space based on appropriate fitness function. A hypothetical numerical experiment is conducted in detail to validate the new method. Results show that the proposed pressure back-analysis method using ANN-GA can effectively determine maximum and minimum horizontal in situ stresses and elastic parameters from borehole pressure values in hydraulic fracturing tests.

INTRODUCTION

Knowledge of in situ stresses and elastic parameters of a subsurface formation is important for performing fracturing operations, drilling operations, oil and/or gas production stimulation, wellbore stability analysis, and coupled geomechanics-reservoir simulation in petroleum engineering.

A number of techniques are available to determine in situ stress orientation including four-arm caliper data, drilling-induced tensile wall fractures, core-based measurements, and wellbore breakouts (Aadnøy, 1990b, Hill et al., 1994, Bell, 1996 and Zoback et al., 2003). Density integrated over depths of overlying layers is often considered to provide good estimation of vertical in situ stress $_v$ (Fjær et al., 2008 and Teichrob et al., 2010). There are various approaches for determination of maximum and minimum horizontal in situ stresses in a formation such as overcoring, strain-relief, inverse solving methods, hydraulic fracturing test, borehole breakouts, acoustic emission, fault plane solution, differential strain analysis, observation of discontinuity states (Haimson, 1978, Zoback et al., 1985, Aadnøy, 1990a, Hareland and Hoberock, 1993, Amadei and Stephansson, 1997 and Seto et al., 1997). Limitations in terms of these approaches have been discussed in detail by Ljunggren et al. (2003), Haimson (1978), and Nicolson and Hunt (2004). An outstanding one is that these methods require precise knowledge of rock elastic parameters when determining horizontal in situ stresses.

Among the aforementioned methods, hydraulic fracturing methods (e.g., mini-frac test, leak-off test) are considered the most effective method for determination of the minimum horizontal in situ stress (White et al., 2002 and Fang and Khaksar, 2011), and the maximum horizontal in situ stress can be determined from breakdown pressure, the minimum horizontal in situ stress and given elastic parameters such as Young's modulus and Poisson's ratio with linear elastic rock behavior assumed for a formation rock (Haimson and Fairhurst, 1969 and Haimson, 1978; Fjær et al., 2008). Haimson (1993) has been making effort to present the detailed history of the method and the complete description of theory, mechanism, and in situ stress derivation.

Previous attempts have been made to use the inverse solving method to determine horizontal in situ stresses on the basis of two conventional hydraulic fracturing criteria (Aadnøy, 1990a and Djurhuus and Aadnoy, 2003). One is the Hubbert and Willis hydraulic fracturing criterion (Hubbert and Willis, 1957). The other is the Haimson and Fairhurst hydraulic fracturing criterion (Haimson and Fairhurst, 1967). With the above criteria, the maximum and minimum horizontal in situ stresses are determined from the borehole pressures in hydraulic fracturing tests based on the following assumptions (Haimson, 1975 and Schmitt and Zoback, 1989): (1) the minimum horizontal in situ stress $_h$ is independently equal to the shut-in pressure; (2) there is a linear elastic relationship between the maximum horizontal in situ stress $_H$ and the breakdown pressure; and (3) the elastic parameters of a formation rock are known.

While in practice, the minimum value between the leak-off pressure, fracture propagation pressure and the shut-in pressure is the closest to the minimum horizontal in situ stress so that it can be determined by these pressures (Schmitt and Zoback, 1989 and Fang and Khaksar, 2011). For determination of the maximum horizontal in situ stress, these elastic parameters are not always available or not accurate due to the lack of high-quality core samples (Ocak and Seker, 2012).

To overcome this difficulty, in this paper, we propose a method that can identify horizontal in situ stresses and elastic parameters at the same time, and therefore the elastic parameters are not needed to be known in advance. This proposed method is a pressure based back analysis method that takes advantage of monitored borehole pressures at multiple points of time in hydraulic fracturing test. Researchers have reported that pumping pressures from hydraulic fracturing tests can be used to determine horizontal in situ stresses and elastic parameters (Hubbert and Willis, 1957; Haimson and Fairhurst, 1967; Garagash and Detournay, 1996). The borehole pressures can be easily monitored with the pressure gauge installed downhole (Cooper et al., 1983). The results generated by this method include multiple parameters such as horizontal in situ stresses of a rock formation and the elastic parameters of the formation rock. Implementation of the inverse analysis follows an indirect back analysis strategy based on an integrated artificial neural network (ANN) and genetic algorithm (GA) model, which has proved effective in multiple parameters identification (Zhang and Yin, 2013, Zhang et al., and Zhang and Yin, 2014).

The rest of this paper is organized as follows. In Section 2, we focus on the governing equations of a forward hydraulic fracturing model. In Section 3, an ANN-GA based back analysis model is described in detail. Section 4 is devoted to giving the needed training and test samples including inputs and outputs for the back analysis model, and to conducting a hypothetical numerical experiment to verify the proposed method. Finally, conclusions are given in Section 5.

FRAMEWORK OF HYDRAULIC FRACTURING MODEL BASED ON DISTINCT ELEMENT METHOD (DEM)

Fracturing in subsurface rock formation involves a strong coupling between fracture propagation, rock deformation, fluid flow, and even heat transfer at great depth (Yin, 2013). Numerous subsurface hydraulic fracturing models have been proposed based on the analytical method, the finite element method, the boundary element method, the discrete element method and other numerical methods. In this work, a simple and convenient 2-D discrete element model is built upon (UDEC) code (Itasca, 2011 and Choi, 2012) to conduct a fully coupled hydraulic-mechanical (HM) analysis and to interpret the relationship between the deformation of intact rock and the hydraulic conductivity in the elastic formation system. The deformation of the fractured rock masses is composed of deformation of intact rock blocks and displacements along and across fractures.

In this model, the constitutive relationship in incremental form is expressed (Lewis and Schrefler, 1998) as

$$d\boldsymbol{\sigma}' = D^e \, d\varepsilon^e + \mathbf{m}\alpha \, dp \tag{1}$$

where $d\boldsymbol{\sigma}'$ is the effective stress increment; $d\varepsilon^e$ is the elastic strain increment; $m=[1, 1, 0]^T$; α is Biot's coefficient; dp is the pore pressure increment. Moreover, D^e is the elastic stress–strain matrix, which can be written as

$$\mathbf{D}^e = \frac{E}{(1+\nu)(1-2\nu)} \begin{bmatrix} 1-\nu & \nu & 0 \\ \nu & 1-\nu & 0 \\ 0 & 0 & \frac{1-2\nu}{2} \end{bmatrix} \tag{2}$$

where E is Young's modulus and v is Poisson's ratio.

In the hydraulic fracturing model, it is assumed that the material model is subject to an isotropic material behavior in the elastic range, and the Coulomb slip is specified for the deformation of the fractures.

In this model, fracturing type during hydraulic fracturing is the edge-to-edge contact. The flow rate in a single fracture of length, l, subject to a pressure difference of dp, is given by the following equation based on the cubic law of flow in fracture (Witherspoon et al., 1980 and Zhang et al., 1999)

$$q = -\frac{a^3}{12\mu}\frac{dp}{l}$$

(3)

where μ is dynamic viscosity. And the contact hydraulic aperture, a, is given by the following relationship:

$$a = a_0 + u$$

(4)

where a_0 is the fracture aperture at zero normal stress, u is the fracture normal displacement, which is related to rock properties and normal stress.

The stress–displacement relation at the contact is assumed to be linear and governed by the normal stiffness k_n and the shear stiffness k_s as

$$k_n = \frac{d\sigma'_n}{du_n}, \quad k_s = \frac{d\sigma'_s}{du_s}$$

(5)

where σ'_n and σ'_s are effective normal stress and effective shear stress, respectively, un and us are normal displacement and shear displacement, respectively.

Fig. 1 shows a typical pressure–time plot with one full pressurization cycle when performing a hydraulic fracture test in a rock formation.

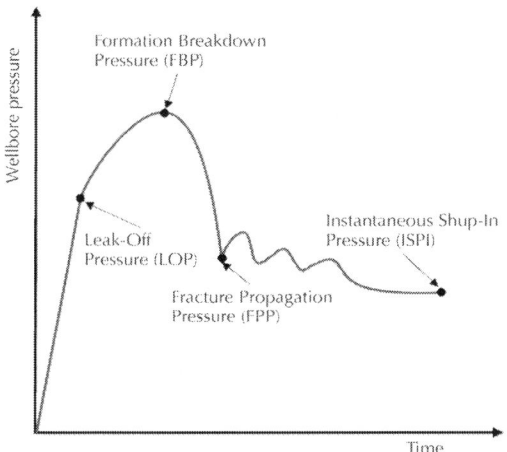

Figure 1: Flowchart of the ANN-GA-based pressure back analysis method.

ANN-GA BASED PRESSURE BACK ANALYSIS METHODOLOGY

In the hybrid ANN-GA model, the ANN model is employed to map the relationship between the maximum and minimum horizontal in situ stresses, elastic parameters and the wellbore pressures, and the detailed description of the ANN model is shown in Appendix A. GA is employed to search the global optimum in a large search space based on the objective function, and the description of GA is shown in Appendix B. Results of back analysis are evaluated by the objective function value and the difference between the predicted pressures, the calculated pressures, and the monitored pressures. The ANN-GA based pressure back analysis method can be described as follows:

Step 1: Build proper ANN by determining initially network type and its algorithm, the number of hidden layers, number of hidden nodes and transfer function, etc.

Step 2: Initialize the weights and biases of the network.

Step 3: Train the initial network. The training process requires a set of examples of proper network behavior – network inputs and target outputs. The weights and biases of network are iteratively adjusted during training.

Step 4: If MSE between the network outputs and the targets is satisfied, or the epoch is reached, the training process will be stopped. Otherwise, repeat Step 3.

Step 5: Check the trained ANN model in terms of MSE performance and data regression results.

Step 6: If both of MSE performance and data regression results are satisfied, the training will end. And then, the best network model topology is saved for GA. Otherwise, go to Step 1.

Step 7: The initializations of GA parameter set including population size, *Npop_size*, maximum generation, *Nmax_gen*, crossover probability, *Pc*, mutation probability, *Pm*, and the range of search space for parameters. In this study, in order to have an effective implementation of GA, the real number encoding method is employed.

Step 8: Generate candidate individuals within the given range of parameters. And then, the initial population is generated based on these candidate individuals. Here, each chromosome (individual) represents an initial solution.

Step 9: Input the generated candidate solutions into the trained and tested ANN model from Step 6. And predicted pressures at the monitoring points are obtained.

Step 10: Use the objective function to evaluate the fitness of current individuals.

Step 11: If all individuals have been evaluated, this model will automatically trace the average fitness and the best individual fitness and go to Step 12. Otherwise, go to Step 9.

Step 12: If the given evolutionary generation is reached, or the best individual is obtained, the algorithm terminates and provides the maximum and minimum horizontal in situ stresses and elastic parameters, as well as the corresponding pressures. Otherwise, go to Step 13.

Step 13: Execute genetic operations, including selection, crossover and mutation. The next generations of selected individuals are obtained based on these genetic operations.

Step 14: Repeat Step 13 until all *Npop_size* new individuals are generated, which are applied as new individuals (offspring).

Step 15: Using the generation of the best parent's individual to replace randomly an individual in the offspring.

Step 16: Take the offspring as parent and go to Step 9.

The entire flow of the ANN-GA based pressure back analysis algorithm is illustrated in Fig. 2.

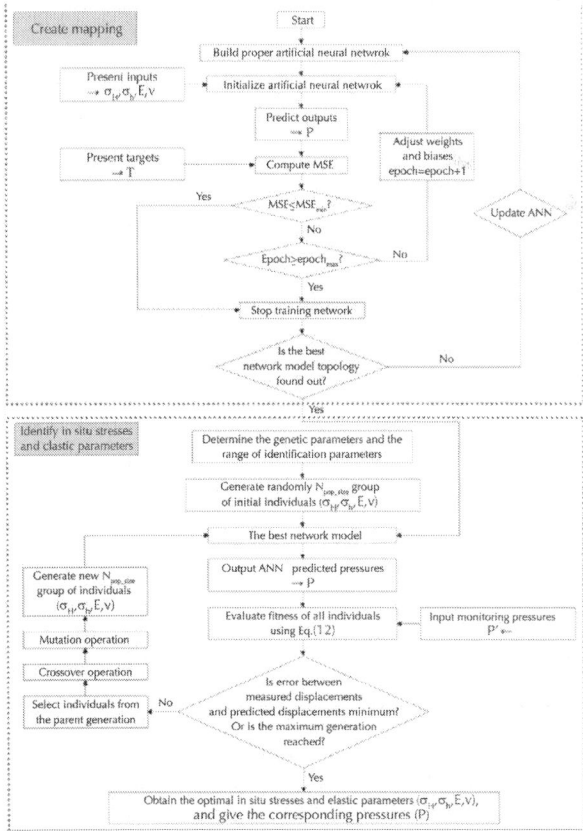

Figure 2: Problem domain geometry and estimated parameters.

NUMERICAL EXPERIMENT

In this section, a numerical experiment is conducted based on the hydraulic fracturing modeling of water injection into a hypothetical deep formation. This is a 2-D plane strain model that is orthogonal to the vertical borehole. Suppose a 44 m×44 m hydraulic area, and 2000 m deep, with a wellbore of 0.2 m in diameter as shown in Fig. 3. In the numerical model, the outer boundary of rock formation is considered to be free boundary.

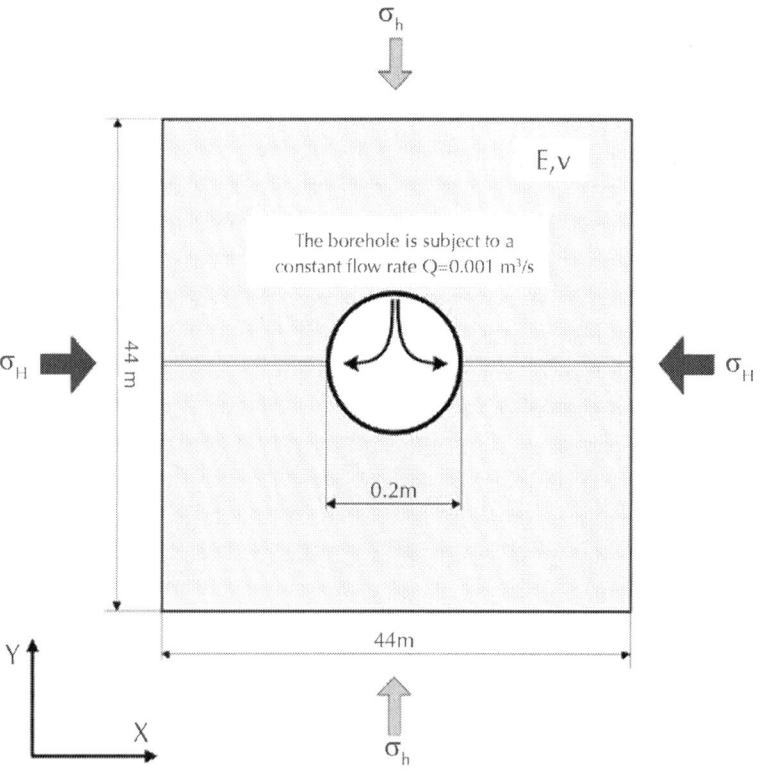

Figure 3: Idealized relationship between wellbore pressure and time during mini-frac test.

To simulate the hydraulic fracture propagating from the wellbore, an existing natural fracture (i.e., joint) is assumed to be parallel to the maximum horizontal in situ stress passing through the center of borehole (seeFig. 3). The assumption is reasonable because (1) the fluid first flows into an existing natural fracture by the pressure in the induced fracture, (2) the hydraulically induced fracture will propagate along the path of least resistance, and (3) the fluid-driven fracture will propagate in a direction orthogonal to the minimum in situ stress. More detailed physical description can be found in references (Mader, 1989 and Olson and Holder, 2012).

The in situ stresses of the fracturing formation are mainly composed of a vertical stress $_v$, and the maximum and minimum horizontal in situ stresses, $_H$ and $_h$. Here $_v$ is orthogonal to the two-dimensional model in vertical borehole so that it is ignored in the plane strain modeling. But the range of $_H$ and $_h$ are usually determined using the ratio of the average horizontal in situ principal stress to the vertical stress. Furthermore, the vertical in situ stress is set equal in this work to 38 MPa, which is subject to the calculated overburden pressure. The physical and mechanical properties of the intact rock and joint used in the numerical experiment are given in Table 1. They are mainly obtained from Choi (2012) and Itasca (2011).

Table 1: Physical and mechanical properties of the intact rock and joint for numerical experiment

Variables	Value
Rock property	
Density (kg/m³)	2600
Joint property	
Normal stiffness (MPa/m)	3×10^5
Shear stiffness (MPa/m)	3×10^5
Cohesion (MPa)	0
Hydraulic fracturing tensile strength (MPa)	15
Friction angle (deg)	45
Permeability factor (1/(Pa s))	83.3
Initial aperture (m)	2×10^{-5}
Maximum aperture (m)	6×10^{-3}

Creation of Training and Testing Samples

In this section, we use the forward model mentioned above to create the needed training and testing samples including inputs and outputs for ANN. Fluid injection at a constant flow rate of 0.001 m³/s is specified for the wellbore, and the injection point is located at the center of model. The joint tensile strength $_{jt}$=13 MPa is used to enforce the zero toughness condition. The compressible flow algorithm is selected and the fluid bulk modulus K_f=100 MPa is specified to simulate an "incompressible" fluid because the value is enough high compared to fluid pressure changes in this model.

As shown in Fig. 4, inputs of training and testing samples are the maximum and minimum horizontal in situ stresses, Young's modulus, and Poisson's ratio. Outputs are specified for the monitored borehole pressures such as leak-off pressure (LOP), formation breakdown pressure (FBP), fracture propagation pressure (FPP), and instantaneous shut-in pressure (ISIP), which are shown in Fig. 1. Therefore, the ANN model mapping the relationship between the maximum and minimum horizontal in situ stresses, elastic parameters and the monitored borehole pressures, is established (see Fig. 4).

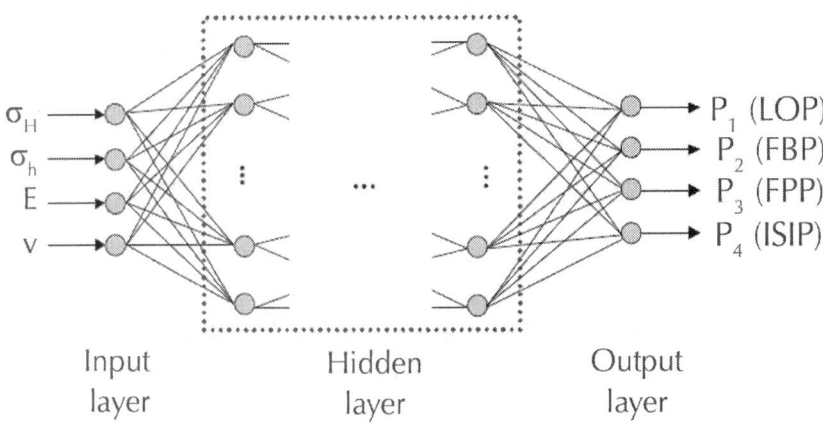

Figure 4: The neural network model.

Finally, 40 training and testing samples with the hypothetical different maximum and minimum horizontal in situ stresses, elastic parameters and the corresponding wellbore pressures for ANN were created. They are listed in Table 2.

Table 2: In situ stress magnitude, elastic parameters and corresponding targeted pressures

No.	In situ stresses and elastic parameters				Targeted pressures (MPa)			
	σ_H	σ_h	E	v	P_1	P_2	P_3	P_4
	(MPa)	(MPa)	($\times 10^3$ MPa)					
Training samples								
1	44.88	34.11	40	0.22	39.778	66.733	40.377	37.482
2	43.83	30.94	42	0.27	35.882	60.069	37.09	32.905
3	43.19	15.74	25	0.33	19.425	25.968	18.394	16.293
4	42.92	20.01	30	0.21	21.183	30.267	22.859	20.486
5	42.31	32.95	40	0.17	37.403	61.538	37.959	34.59
6	41.89	28.48	42	0.3	33.423	54.845	34.099	30.028
7	40.95	17.76	26	0.26	21.578	27.741	20.437	18.33
8	40.41	31.27	47	0.26	37.074	60.791	37.452	33.623
9	39.79	26.96	38	0.19	32.056	43.779	33.969	28.521
10	39.24	11.34	28	0.22	14.387	21.38	13.928	11.916
11	38.88	29.85	41	0.28	35.089	57.713	35.694	31.699
12	38.09	18.93	35	0.27	23.467	30.347	22.057	19.47
13	37.55	27.41	27	0.25	31.439	51.838	32.584	28.979
14	36.49	14.87	37	0.24	19.223	27.261	18.048	15.485
15	35.52	25.14	39	0.23	30.255	39.013	28.688	26.052
16	34.61	12.76	24	0.32	16.305	23.295	15.414	13.331
17	33.83	28.91	42	0.29	34.409	55.727	34.523	30.596
18	33.17	17.86	32	0.28	22.06	28.912	20.846	18.474
19	32.49	10.34	36	0.18	13.789	20.18	13.122	10.948
20	31.51	24.86	21	0.22	28.983	45.886	29.211	25.937
21	31.07	16.87	24	0.28	20.342	26.571	19.426	17.392
22	30.46	22.67	35	0.26	27.216	35.282	25.95	23.289

23	29.53	25.14	40	0.17	29.886	39.043	28.714	26.014
24	28.42	13.55	28	0.24	17.67	24.647	16.346	14.127
25	27.79	20.17	32	0.21	24.602	30.77	23.131	20.661
26	27.18	15.81	26	0.31	19.189	26.153	18.522	16.371
27	26.92	13.05	43	0.25	17.573	24.725	16.366	13.71
28	26.27	20.58	27	0.26	24.947	30.101	23.283	21.048
29	25.73	21.34	38	0.19	26.112	33.84	24.708	21.888
30	25.12	11.05	21	0.25	14.721	20.638	13.348	11.579
Testing samples								
31	43.14	33.17	40	0.25	37.941	62.638	38.494	34.884
32	41.38	20.42	32	0.27	24.674	31.735	23.49	20.921
33	39.79	24.51	42	0.21	29.1	38.351	28.134	25.292
34	37.53	12.94	28	0.32	17.039	24.081	15.802	13.541
35	35.29	27.81	38	0.19	32.43	52.567	32.982	29.162
36	33.63	18.56	30	0.24	22.887	29.136	21.431	19.078
37	32.07	11.93	24	0.28	15.405	22.233	14.451	12.487
38	29.69	21.45	42	0.18	25.693	34.547	24.928	22.672
39	27.62	16.58	27	0.22	20.192	26.886	19.239	17.129

Verification of ANN-GA Based Pressure Back Analysis Method

In this section, the proposed back analysis method is verified against the data generated in Section 4.1. The ANN model representing the relationship between the maximum and minimum horizontal in situ stresses, Young's modulus, Poisson's ratio and the borehole pressures, is trained and tested with the training and testing samples including inputs and outputs (see Table 2). Next, GA is applied to search the global optima (e.g., σ_H, σ_h, E, v) in a large search space based on the objective function.

For the ANN model, MSE, R-value, and cross plot on targets versus predicted values are used as diagnostic tools to check if the ANN model can map effectively the relationship between the maximum and minimum horizontal in situ stresses, elastic parameters and the borehole pressures.

Variations of MSE for training, validation and testing with iterations are shown in Fig. 5. After 20 iterations, the training of the ANN model terminates because the validation error starts increasing from 0.0084 to 0.0086. It can be seen that the final training set error of 0.00037 is very close to the true value of 0 and the validation set error and the testing set error have similar characteristics, which indicates that the training process of ANN is satisfactory.

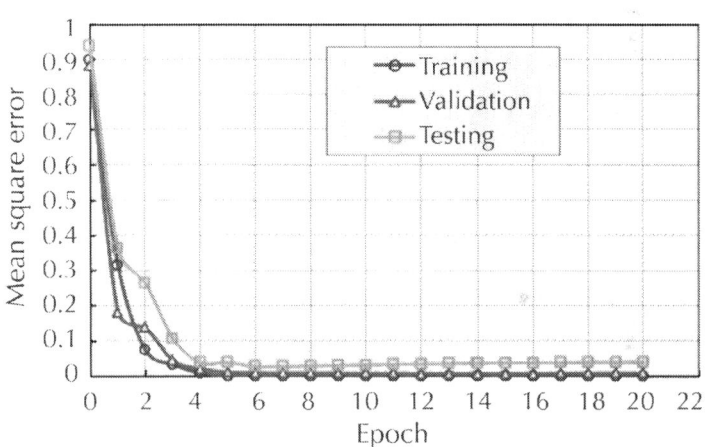

Figure 5: Variation curves of the mean square error with epoch.

Fig. 6 shows the linear regression between targets and network outputs and the R-values for training, validation, testing and all samples. Scatter plots for all the four phases show good correlations between targets and predicted values. All the R-values are greater than 0.95, which demonstrates that the ANN model performs excellently (Yilmaz and Yuksek, 2008).

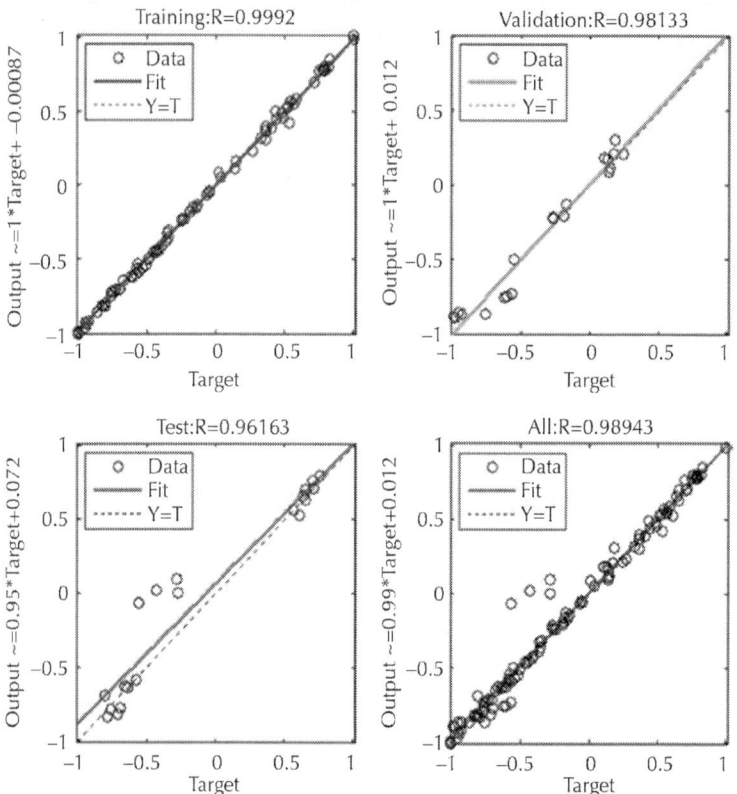

Figure 6: The linear regressions between the network outputs and the corresponding targets.

Fig. 7 shows the cross plot on targeted pressure values versus ANN-predicted pressure values for training and testing samples. As shown in Fig. 7, the ANN-predicted pressures are in agreement with the targeted pressures.

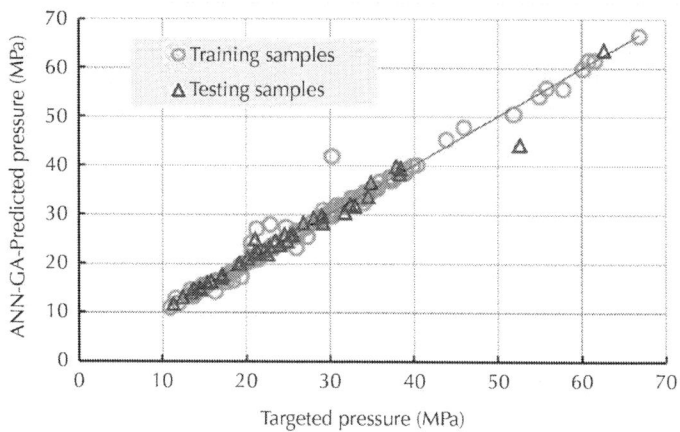

Figure 7: Cross plot on targeted and ANN-GA-predicted pressures.

The above-mentioned results demonstrate that ANN can accurately map the relationship between the maximum and minimum horizontal in situ stresses, elastic parameters and the wellbore pressures to provide a good objective function for GA to search.

In the pressure back analysis, parameters of GA are set as follows: maximum generation, N_{max_gen}=1000, population size, N_{pop_size}=80, crossover probability, P_c=0.7, and mutation probability, P_m=0.1.

Furthermore, the ranges of parameters determined by a hybrid ANN-GA model are set as follows: the maximum horizontal in situ stress, $_H$: 25.00–45.00 MPa, the minimum horizontal in situ stress, $_h$: 10.00–35.00 MPa; for elastic properties, Young's modulus, E: 20–50×10³ MPa, Poisson's ratio, v: 0.15–0.35.

Suppose the monitored wellbore pressure values obtained by a hydraulic fracturing test are as follows: LOP, P'_1=19.572 MPa; FBP, P'_2=27.362 MPa; FPP, P'_3=18.296 MPa; and ISIP, P'_4=15.637 MPa. And these are substituted into GA to set up the objective function by combining with the corresponding wellbore pressure values predicted by the ANN model.

After genetic operation of 1000 generations of evolution, identified results with the ANN-GA based pressure back analysis method are shown in Table 3. The corresponding borehole pressures predicted by the ANN model are shown in Table 4 for comparison. Moreover, by inputting the identified parameters into the UDEC hydraulic fracturing model, the calculated results on wellbore pressures are also shown in Table 4 for comparison.

Table 3: Comparison of in situ stress between the ANN-GA model and H–W criterion solution

	σ_H	σ_h	E	v
	(MPa)	(MPa)	($\times 10^3$ MPa)	
ANN-GA model	35.83	14.9681	39.92	0.26
H–W criterion	34.549	15.637	–	–
Absolute error	1.281	0.6689	–	–
Relative error (%)	3.5	4.47	–	–

Table 4: Monitored, predicted and calculated pressure values and their errors

No.	Pressure values (MPa)			Absolute error (MPa)		Relative error (%)	
	Moni-tored	ANN-GA	UDEC	Monitored and ANN-GA	Moni-tored and UDEC	Monitored and ANN-GA	Moni-tored and UDEC
P_1(LOP)	19.572	18.9659	19.639	0.6061	0.067	3.097	0.342
P_2(FBP)	27.362	27.6301	27.428	0.2681	0.066	0.98	0.241
P_3(FPP)	18.296	18.2988	18.314	0.0028	0.018	0.015	0.098
P_4(ISIP)	15.637	15.638	15.614	0.001	0.023	0.006	0.147

For comparison and verification, we select the solution of the modified Hubbert and Willis (H–W) criterion (Schmitt and Zoback, 1989) for the particular case which is similar to ours. Schmitt and Zoback have developed an analytical solution for the maximum in situ stress without considering the effect of Young's modulus and Poisson's ratio. The maximum in situ stress can be expressed as (Schmitt and Zoback, 1989)

$$\sigma_H = 3\sigma_h - P'_2 + T \tag{6}$$

The H–W criterion solution is done with the parameters of the mentioned above case: $\sigma_h = P'_4 = 15.637$ MPa, $P'_2 = 27.362$ MPa, $T = 15.00$ MPa. And the comparison is shown in Table 3. The results by the ANN-GA based pressure back analysis method show consistency with that of analytical solution.

Fig. 8 shows variations of objective function value with generations for both the average individual fitness and the best individual fitness. It can be seen that the GA model keeps the diversity of the individual during the adaptive heuristic search process and has a satisfactory fitness level ultimately. The final average fitness and the final best fitness are 0.0153 and 0.0152, respectively, which shows the GA can find the global optimum in a large search space based on the well-established objective function.

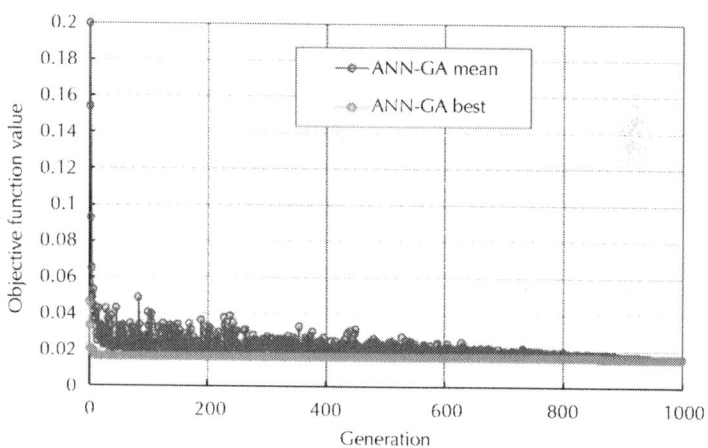

Figure 8: Variations of objective function value with generations.

Fig. 9 shows the comparison of pressure values from wellbore at four monitoring points. It can be seen that the ANN-predicted and UDEC-calculated wellbore pressures are in good agreement with the monitored borehole pressures.

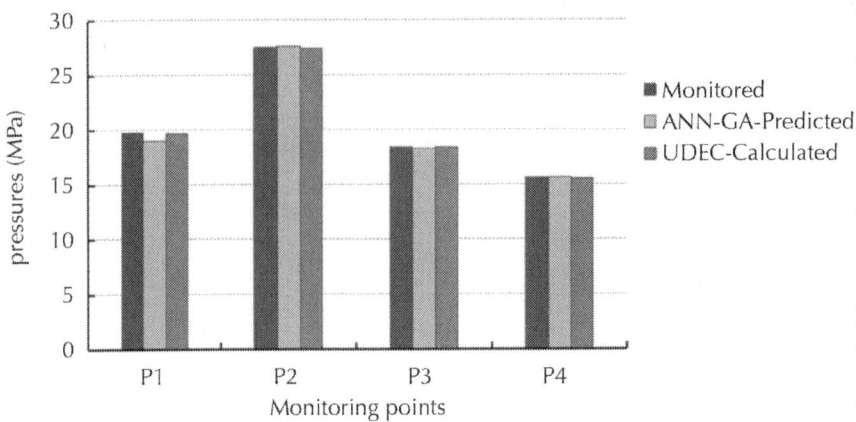

Figure 9: Comparison on monitored, predicted and calculated pressures at monitoring points.

Table 4 shows the specific wellbore pressure values and the absolute and relative errors among them including ANN-predicted, UDEC-calculated, and monitored pressures. It can be seen that all the absolute errors are less than 1.0 MPa, and all the relative errors are less than 5%, as expected. The results from the ANN model and the UDEC model show high consistency with that of the monitored pressures at the monitoring points.

Results from two important evaluation criteria demonstrate that the proposed hybrid ANN-GA model not only can accurately identify the maximum and minimum horizontal in situ stresses and elastic parameters from wellbore pressure values, but can also predict the borehole pressures in hydraulic fracturing.

CONCLUSIONS

In this paper, an ANN-GA based pressure back analysis method is developed to determine the maximum and minimum horizontal in situ stresses, Young's modulus, and Poisson's ratio from the monitored pressures in hydraulic fracturing tests for specified stress conditions given geomechanical properties of the rocks and the fluid compressibility. This method integrates ANN, GA and numerical analysis. The trained ANN is used to represent the correlation between the in situ stresses, elastic parameters and the wellbore pressures and to provide the objective function for GA. GA is then employed to search the global optimum in the large search space based on the objective function.

With a hybrid ANN-GA model, a hypothetical numerical experiment on hydraulic fracturing is investigated in order to determine the maximum and minimum horizontal in situ stresses, elastic parameters. Results of MSE, R-values and cross plot demonstrate clearly that the ANN model has an excellent description for correlation between the maximum and minimum horizontal in situ stresses, elastic parameters and the borehole pressures. Results of the objective function values and the comparison between the ANN-GA predictions, simulation results and monitored pressures indicate that the ANN-GA based pressure back analysis method can be used to determine the maximum and minimum horizontal in situ stresses and elastic parameters from wellbore pressure in hydraulic fracturing tests.

Appendix A. Artificial Neural Network

The artificial neural network (ANN) was originally developed in the 1940s by McCulloch and Pitts (1943). An ANN is composed of a large number of nodes and weights between nodes. Each node besides the input nodes is a processing element (or neuron) by using an activation function. Network output changes according to difference of connected type, weight and activation function. Network itself is a type of natural algorithm or logical expression. Thus, neural networks can be considered to be the modeling tools of linear and/or nonlinear statistical data. A single artificial neuron with a node threshold, b, connection weights, $w_i (i=1, 2, 3, \ldots, n)$ and a transfer function $p^{(k)}=f(x^{(j)})$ is shown in Fig. A1. For each pattern j ($j=1, 2, 3, \ldots, m$), all patterns can be expressed in matrix notation as (Nikravesh et al., 2003)

$$
\begin{bmatrix} x^{(1)} \\ x^{(2)} \\ \vdots \\ x^{(m)} \end{bmatrix} = \begin{bmatrix} x_1^{(1)} & x_2^{(1)} & \cdots & x_n^{(1)} & 1 \\ x_1^{(2)} & x_2^{(2)} & \cdots & x_n^{(2)} & 1 \\ \vdots & \vdots & \ddots & \vdots & \vdots \\ x_1^{(m)} & x_2^{(m)} & \cdots & x_n^{(m)} & 1 \end{bmatrix} \begin{bmatrix} w_1 \\ w_2 \\ \vdots \\ w_n \\ b \end{bmatrix}
$$

(A.1)

or in a more compact form for j=1, 2, 3, ..., m

$$
x^{(j)} = x_1^{(j)} w_1 + x_2^{(j)} w_2 + \cdots + x_n^{(j)} w_n + b
$$

(A.2)

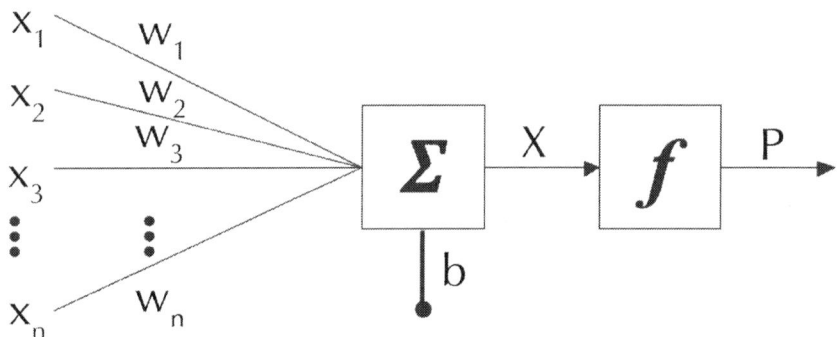

Figure A1: Schematic of a single artificial neuron.

Next, predicted pressures can be given by a transfer function:

P=ANN(X)=f(X) (A.3)

$$
X = (x^{(1)}, x^{(2)}, \ldots, x^{(m)})
$$

P=(p$^{(1)}$,p$^{(2)}$,...,p$^{(k)}$)

In this mathematical model, $x^{(j)}$ (j=1, 2, 3, ..., m) represents inputs with weights, thresholds; $p^{(k)}$ (k=1, 2, 3, ..., M) stands for predicted outputs.

Both mean square error (MSE) and correlation coefficient (R-value) are used to evaluate the ANN model performance. MSE is defined below as the average sum of squares of the difference between predicted pressures and targeted pressures (Shoo and Ray, 2006):

$$MSE = \frac{1}{N} \sum_{r=1}^{N} (P_r - T_r)^2$$

(A.4)

where N is the number of samples, P_r and T_r are the ANN-predicted pressures and the targeted pressures, respectively.

R-value is obtained by performing a linear regression between the targeted pressures and the predicted pressures and can be expressed as (Sahoo and Ray, 2006)

$$R = \frac{\sum_{r=1}^{N} t_r p_r}{\sqrt{\sum_{r=1}^{N} t_r^2} \sqrt{\sum_{r=1}^{N} p_r^2}}$$

(A.5)

where N is the number of sample, $t_r = T_r - \overline{T}$, $p_r = P_r - \overline{P}$.

Once the ANN model performance is satisfactory, it will be applied to represent the relationship between inputs (e.g., in situ stresses, elastic parameters) and outputs (e.g., leak-off pressure, breakdown pressure, fracture propagation pressure, shut-in pressure) for pressure prediction.

Appendix B. Genetic Algorithm

The genetic algorithm (GA) was originally introduced in the early 1960s by John Holland (1975). GA is a global search and optimization technique based on some principles from evolution theory. The technique starts with a set of solutions to the problem, this set of solutions is called the population, and each individual in the population is called a chromosome. These chromosomes generated by successive iterations are evaluated based on the objective function. A roulette wheel selection is adopted to implement the selection operator of GA to determine which chromosomes are selected as parents. And parents create next generations, new chromosomes, also called offspring through crossover and mutation operations. Now the $S_g=[s_{g1}, s_{g2}, ..., s_gN]$ is used to represent chromosome g in the population $g=1, 2, ..., N_{pop\_size}$. The fitness is $eval(S_g)=f(S_g)$ for each chromosome S_i. And then, total fitness can be calculated for the population by (Gen and Cheng, 1997)

$$S = \sum_{g=1}^{N_{pop\_size}} eval(S_g)$$
(B.1)

Selection probability for each chromosome S_g can be expressed as (Gen and Cheng, 1997)

$$F_g = \frac{eval(S_g)}{S}$$
(B.2)

As in any traditional approaches for the back analysis, an objective function is necessary to be defined when GA is used to search the maximum and minimum horizontal in situ stresses and elastic parameters in a large search space. The objective function can be defined as

$$fitness = \min \left(\frac{1}{M} \sum_{k=1}^{M} \left(|P_k - P'_k| \right) \right)$$
(B.3)

where P_k is the predicted pressure of the kth monitoring point. P'_k is the monitored pressure of the kth monitoring point. M is the number of monitoring point.

After establishing the objective function, GA can rapidly search the global optimum (e.g., σ_H, σ_h, E, v) on the basis of the above objective function.

REFERENCES

1.   Aadnøy, B.S., 1990a.Inversion technique to determine the in-situstress field from fracturing data.J.Pet.Sci.Eng.4 (2), 127–141.

2. Aadnøy, B.S., 1990b.In-situ stress direction from borehole fracture traces.J.Pet.Sci. Eng. 4(2), 143–153.

3. Amadei, B.,Stephansson,O.,1997.Rock Stress and its Measurement.Chapman & Hall, London,UK.

4. Bell, J.S.,1996.Petro Geoscience1.In Situ Stresses in Sedimentary Rocks(part1): Measurement Techniques.Geosci. Canada23(2),85–100.

5. Choi, S.O.,2012.Interpretation of shut-in pressure in hydrofracturing pressure– time records using numerical modeling.Int.J.Rock Mech.Min.Sci.50(2012), 29–37.

6. Cooper,G.D.,Nelson,S.G.,Schopper,M.D.Improving fracturing design through the use of a on-site computer system. SPE 12063 presented at the 1983 SPE Annual Technical Conference & Exhibition,San Francisco,CA.

7. Djurhuus, J.,Aadnoy,B.S.,2003.In situ stress state from inversion of fracturing data from oil wells and borehole image logs.J.Pet.Sci.Eng.38(3–4), 121–130.

8. Fang,Z.,Khaksar,A.,2011.Complexity of minifrac tests and implications for in-situ horizontal stresses in coalbed methane reservoirs.IPTC 14630,1–13.

9. Fjær, E.,Holt,R.M.,Horsrud,P.,Raaen,A.M.,Risnes,R.,2008. Petroleum Related Rock Mechanics,2nded.Elsevier B.V.,UK.

10. Garagash, D.,Detournay,E.,1996.Influence of pressurization rate on borehole break down pressure in impermeable rocks. ARMA96–1075,1075–1080.

11. Gen, M.,Cheng,R.,1997.Genetic Algorithms and Engineering Design.John Wiley & Sons, NewYork,USA.

12. Haimson, B.C.,Fairhurst,C.,1967.Initiation and extension of hydraulic fractures in rocks.SPEJ.7(3),310–318.

13. Haimson, B.C.,Fairhurst,C.,1969.In-situ stress determination at great depth by means of hydraulic fracturing.In:The 11th U.S.Symposium on Rock Mechanics (USRMS). Berkeley,CA,pp.559–584.

14.   Haimson, B.C.,1975.Deepin-situ stress measurements by hydrofracturing.Tecto- nophysics 29(1975),41–47.

15.   Haimson, B.C.,1978.The hydrofracturing stress measuring method and recent field results. Int.J.Rock Mech.Min.Sci. Geomech.Abstr.15,167–178.

16.   Haimson, B.C.,1993.The hydraulic fracturing method of stress measurement: theory and practice.In:Hudson,J .A.(Ed.),Comprehensive Rock Engineering,3. Pergamon Press,Oxford,pp.395–412.

17.   Hareland, G.,Hoberock,L.L.,1993.Use of drilling parameters to predict in-sitestress bounds. SPE/IADC25727,457–471.

18.   Hill,   R.E.,Peterson,R.E.,Warpinski,N.R.,Teufel,L.W.,Aslaks on,J.K.,1994.Techniques for determining subsurface stress direction and assessing hydraulic fracture azimuth.SPE 29192,305–320.

19.   Holland, J.H.,1975.Adaptation in Natural and Artificial Systems.University of Michigan Press,AnnArbor,MI,USA.

20.   Hubbert, M.K.,Willis,D.G.,1957.Mechanics of hydraulic fracturing.Pet.Trans., AIME 210,153–168.

21.   Itasca,   2011.Universal Distinct Element Code (UDEC)   Version5.0.Itasca   Consulting   Group Inc,Minneapolis,Minnesota,USA.

22.   Lewis,R.W.,Schrefler, B.A.,1998.The Finite Element Method in the Staticand Dynamic Deformation and Consolidation of Porous Media,John Wiley and Sons.

23.   Ljunggren,   C.,Chang,Y.,Janson,T.,Christiansson,R.,2003.An over view of rock stress measurement methods.Int.J.Rock Mech.Min.Sci.40(7),975–989.

24.   Mader, D.,1989.Hydraulic Proppant Fracturing and Gravel Packing.Elsevier Science Ltd., NewYork.

25.   McCulloch, W.S.,Pitts,W.H.,1943.A logical calculus of the ideas immanent in nervous activity.Bull.Math.Biol.5(4),115–133.

26. Nicolson,J.P.W.,Hunt,S.P.Distinct element analysis of borehole instability in fractured petroleum reservoir seal formation,SPE 88610.In:Proceedings of the 2004 SPE Asia Pacific Oil and Gas Conference and Exhibition (APOGCE),Perth, Australia,18–20 October2004.

27. Nikravesh,M.,Aminzadeh,F.,Zadeh,L.A.,2003.Soft Computing and Intelligent Data Analysis in Oil Exploration. Elsevier Science B.V.,Boston.

28. Ocak, I.,Seker,S.E.,2012.Estimation of elastic modulus of intact rocks by artificial neural network.Rock Mech.Rock Eng.45 (2012),1047–1054.

29. Olson, J.E.,Holder,B.J.Examining hydraulic fracture:fracture:natural fracture interaction in hydrostone block experiments.SPE 152618 presented at the SPE Hydraulic Fracturing Technology Conference,The Woodlands, 6-8 February 2012.

30. Sahoo, G.B.,Ray,C.,2006.Flow forecasting for a Hawaiian stream using rating curves and neural networks.J.Hydrol.317(1),63–80.

31. Schmitt, D.R.,Zoback,M.D.,1989.Poroelastic effects in the determination of the maximum horizontal principal stress inhydraulic fracturing tests a proposed breakdown equation employing a modified effective stress relation for tensile failure. Int.J.Rock Mech.Min.Sci.Geomech.Abstr.26 (6), 499–506.

32. Seto, M.,Utagawa,M.,Katsuyama,K.,Nag,D.K.,Vutukuri,V .S.,1997.In situ stress determination by acoustic emission technique.Int.J.Rock Mech.Min.Sci. 34 (3),1–16.

33. Teichrob,R.,Kustamsi,A.,Hareland,G.,Odiegwu,U.Estimating in situ stress magnitudes and orientations in an Albertan field in Western Canada, ARMA- 10–224. Presented at the 44thU.S.Rock Mechanics Symposium, 27–30 June, 2010,Salt Lake City,UT.

34. White, A.J.,Traugott,M.O.,Swarbrick,R.E.,2002.The use of leak-off tests as means of predicting minimum in-situ stress. Pet.Geosci.8,189–193.

35. Witherspoon, P.A.,Wang,J.S.Y.,Iwai,K.,Gale,J.E.,1980.Validity of cubic law for fluid flow in a deformable rock fracture.Water Resour.Res.16 (6),1016–1024.

36. Yilmaz, I.,Yuksek,A.G.,2008.An example of artificial neural network (ANN) application for indirect estimation of rock parameters.Rock Mech.Rock Eng. 41 (5),781–795.

37. Yin, S.,2013.Numerical analysis of thermal fracturing in subsurface cold water injection by finite element methods. Int.J.Numer.Anal.Meth.Geomech. 37(15), 2523–2538.

38. Zhang, S.,Yin,S.,2013.Reservoir geomechanical parameters identification based on ground surface movements.Acta Geotech.8(3), 279–292.

39. Zhang, S.,Yin,S.,2014.Determination of horizontal in-situ stresses and natural fracture properties from well boredeformation.Int.J.Oil Gas Coal Technol.7(1), 1–28.

40. Zhang, S.,Yin,S.,Yuan,Y.,2013.Estimation of fracture stiffness,in situ stresses,and elastic parameters of naturally fractured geothermal reservoirs.Int.J.Geomech. doi: 10.1061/ (ASCE) GM.1943-5622.0000380,04014033.

41. Zhang, X.,Last,N.,Powrie,W.,Harkness,R.,1999.Numerical modeling of wellbore Behavior in fractured rock masses.J.Pet. Sci.Eng.23 (2),95–115.

42. Zoback, M.D.,Barton,C.A.,Brudy,M.,Castillo,D.A.,Finkbeiner ,T.,Grollimund,B.R., Moos, D.B.,Peska,P.,Ward,C.D.,Wiprut, D.J.,2003.Determination of stress orientation and magnitude in deep wells.Int.J.Rock Mech.Min.Sci.40 (7), 1049–1076.

43. Zoback, M.D., Moors, D., Mastin,L.,1985.Well bore breakouts and in situ stress. J. Geophys.Res.90 (B7), 5523–5530.

Microstructure and Fabric Development in Ice: Lessons Learned from in Situ Experiments and Implications for Understanding Rock Evolution

Christopher J.L. Wilson[a], Mark Peternell[b], Sandra Piazolo[c], and Vladimir Luzin[d]

[a]School of Geosciences, Monash University, Clayton, Victoria 3800, Australia

[b]Institute of Geosciences, University of Mainz, 55128 Mainz, Germany

[c]Australian Research Council Centre of Excellence for Core to Crust Fluid Systems/GEMOC, Department of Earth and Planetary Sciences, Macquarie University, NSW 2109, Australia

[d]ANSTO Locked Bag 2001, Kirrawee DC, Lucas Heights, NSW 2232, Australia

ABSTRACT

In this contribution we present a review of the evolution of microstructures and fabric in ice. Based on the review we show the potential use of ice as an analogue for rocks by considering selected examples that can be related to quartz-rich rocks. Advances in our understanding of the plasticity of ice have come from experimental investigations that clearly show that plastic deformation of polycrystalline ice is initially produced by basal slip. Interaction of dislocations play an essential role for dynamic recrystallization processes involving grain nucleation and grain-boundary migration during the steady-state flow of ice. To support this review we describe deformation in polycrystalline 'standard' water-ice and natural-ice samples, summarize other experiments involving bulk samples and use in situ plane-strain deformation experiments to illustrate the link between microstructure and fabric evolution, rheological response and dominant processes. Most terrestrial ice masses deform at low shear stresses by grain-size-insensitive creep with a stress exponent ($n \leq 3$). However, from experimental observations it is shown that the distribution of plastic activity producing the microstructure and fabric is initially dominated by grain-boundary migration during hardening (primary creep), followed by dynamic recrystallization during transient creep (secondary creep) involving new grain nucleation, with further cycles of grain growth and nucleation resulting in near steady-state creep (tertiary creep). The microstructural transitions and inferred mechanism changes are a function of local and bulk variations in strain energy (i.e. dislocation densities) with surface grain-boundary energy being secondary, except in the case of static annealing. As there is a clear correspondence between the rheology of ice and the high-temperature deformation dislocation creep regime of polycrystalline quartz, we suggest that lessons learnt from ice deformation can be used to interpret polycrystalline quartz deformation. Different to quartz, ice allows experimental investigations at close to natural strain rate, and through in-situ experiments offers the opportunity to study the dynamic link between microstructural development, rheology and the identification of the dominant processes.

INTRODUCTION

Deformation in ice occurs on a variety of scales and has been used by numerous workers as an analogue for processes that contribute to rock deformation. The reason for this is that glacier ice and rock at high metamorphic grade deform according to the same non-linear flow laws. Ice is therefore an ideal analogue as it contains folds, faults, boudinage structures and shear zones that can be observed at both the mesoscale and macroscale (Nye, 1953; Hambrey, 1977; Marmo and Wilson, 1998) and has been used successfully as a rock analogue (Wilson, 1981, 1983). In contrast to rocks, in glaciers it is possible to measure directly the strain rate associated with the development of structures in glaciers. By integrating strain rate measurements with constitutive flow laws and material properties one can determine the distribution and history of the stress. It is generally agreed that upon application of a constant stress or constant strain rate, a sample of polycrystalline ice with a random fabric will show an initial elastic deformation followed by a stage of decelerating creep rate, and finally a stage of accelerating creep rate (Budd and Jacka, 1989). The acceleration may be a result of; (1) dislocation multiplication and rearrangement on a submicroscopic scale (Montagnat and Duval, 2000), (2) recrystallization and the development of a non-random fabric (Wilson, 1986); and/or (3) formation of microcracks and fractures (Golding et al., 2010, 2012).

As in many crystalline materials, such as rocks and metals, deformation in ice crystals mostly occurs by crystal-plastic mechanisms that give rise to dynamic recrystallization, that involve grain-boundary migration and dynamic recrystallization (Duval et al., 1983). Available evidence indicates that the driving force for flow of ice during natural deformation is caused mainly by temperature and stress differences (Ma et al., 2010) induced by gravitational forces acting on the sloping ice body (Donoghue and Jacka, 2009). Major ice sheets experience cycles of long-range stress fields that lead to plastic deformation (Taupin et al., 2008) or cycles of elastic stress build-up during interseismic periods followed by rapid stress drops during crevasse formation (Van der Veen, 1998). Furthermore, a particular volume of ice will be subject to changing temperature, depth and deviatoric stress conditions, where temperature and depth will change with time. During deformation of natural ice, strain rate increases as temperature increases so that microstructures formed under steadily increasing deviatoric stress are superimposed. These complications make it exceptionally difficult to extract information from natural-ice microstructures and relate it to the evolution of crystallographic preferred orientations, namely c-axis fabrics (Fig. 1a–b), and also to a strain history. Experiments show that creep strength in natural ice is also strain rate dependent (Barnes et al., 1971); therefore it must respond to these stress cycles and will not be steady-state. Change in the strain rates accommodated by different creep mechanisms that accompany stress cycling, control creep mechanics (rheology) and resultant microstructures. Most laboratory ice-creep experiments aim to achieve steady-state tertiary creep (Fig. 2) and there is a lack of experiments involving high stress/strain rate deformation that is followed by static annealing that would give some insight into the more complicated stress-temperature histories likely in a glacier (Paterson, 1994).

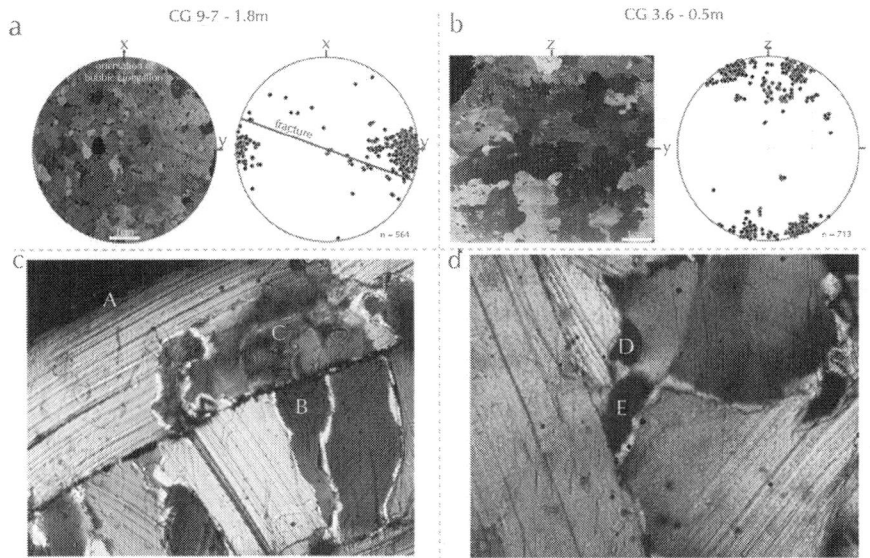

Figure 1: Examples of grain microstructures and c-axis variations in polar ice of the Sørsdal Glacier and experimentally deformed ice. (a) Section cut perpendicular to ice layering in Sørsdal Glacier (1.8 m in CG 9.7; see Wilson and Peternell, 2012) illustrating the interlocking nature of the grains that overgrow air bubbles and cut by a later fracture. The ice crystallographic preferred orientation is characterized by a single maximum c-axis crystal orientation pattern perpendicular to the air bubble alignment (x). (b) Ice cut perpendicular to the flow layering and air bubble alignment in Sørsdal Glacier (0.5 m in CG 3.6; see Wilson and Peternell, 2012), illustrating undulose extinction and small-circle girdle of c-axes (c) Basal slip lamellae that reflect the basal slip in anisotropic ice grains (e.g. grain A) in a 0.7 mm thick sheet of ice deformed deformed at −5 °C and strain rate of 8.7×10^{-7} s^{-1} during a 2D in situexperimental deformation. Other smaller grains (e.g. grain B) have undergone grain-boundary migration. Overgrowing grain A is a set of new grains (C). (d) Experimentally deformed ice with nucleation of new grains (D and E) along grain boundaries, between grains that display slip lines produced by basal slip in differently oriented grains.

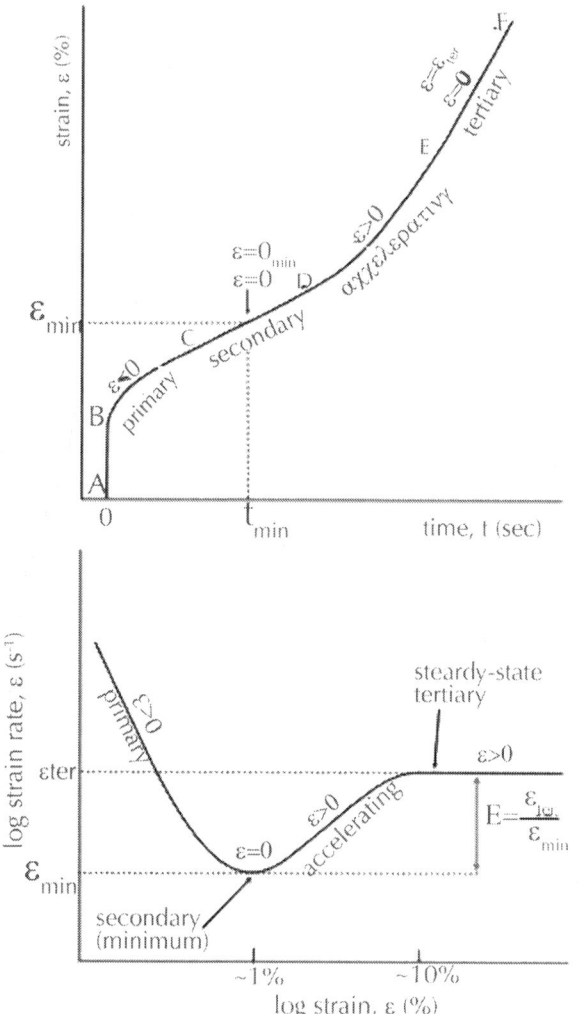

Figure 2: An idealized creep curve for ice shows the various stages of strain, versus time, t as follows A–B: the initial elastic strain, B–C, C–D, D–E and E–F the primary, secondary, accelerating and tertiary stages. (b) The corresponding log strain rate versus log strain plot shows the same features with enhanced flow during tertiary creep resulting from the activation of recovery and recrystallization processes at strains exceeding ~1%. The enhancement factor (E = ter/min) provides a simple parameterization of tertiary creep rates that glaciologists use as an input into ice sheet dynamic models (e.g. Budd and Jacka, 1989).

Both ice core data and ice-deformation laboratory experiments provide very good systems to study the deformation heterogeneity development and dynamic recrystallization mechanisms for highly anisotropic materials. As such they constitute a valuable data set to validate numerical modelling approaches for polycrystal mechanical behaviour (Lebensohn et al., 2007; Montagnat et al., 2011; Griera et al., 2013). Major advances in the characterization of the heterogeneous deformation of ice has been provided by microstructural observations derived from experiments in transmitted light, which permit direct observation of grain-scale processes during dynamic and static recrystallization (Wilson, 1986). Similar advances are found in the geological literature where there is a range of experiments that attempt to quantify microstructural and rheological evolution of rocks (Means, 1980). It is the microstructure of ice and its response to a stress regime that holds the key to understanding the macroscopic rheological behaviour of glaciers and ice sheets and the nature of their flow mechanisms. Thus, to understand the behaviour on a macroscale it is essential to understand substructure dynamics and how this exerts an influence on the behaviour of ice sheets and glaciers. Similarly post-deformational annealing is of particular interest, as this is often the last process to affect a rock, occurring when pressure and deviatoric stress are removed but the system still retains or is subjected to a high temperature. If we understand this process then we can potentially see through to the previous deformation conditions. As in quartz-rich rocks (e.g. Hirth and Tullis, 1992; Trepmann et al., 2007) the creep strength in ice is strain rate dependent (Budd and Jacka, 1989) and by focussing on the dominant mechanisms in terrestrial ice we can use it as an analogue for the microfabric development in crustal rocks deformed at high temperatures. For example, research on ice deformation shows that changes in strain rates can be accommodated by different creep mechanisms including grain-size-sensitive mechanisms that accompany stress cycling and control creep mechanics (Duval et al., 1983). Coarser grain size allows access to more grain-size-insensitive (dislocation) dominated regions where grain size reduction occurs by dynamic recrystallization. Finer grains sizes allow faster grain-size-sensitive (diffusion and grain-boundary sliding) dominated regimes (Goldsby, 2006) but are more problematic at higher temperatures. Because grain growth, accompanied by grain-boundary migration, will alter rheology and strain rate, over the duration of the deformation; this is similar to observations in monomineralic crustal rocks such as marble and quartzite. To a certain extent ice can also simulate the processes recognized in polymineralic rocks through the presence of air bubbles and debris such as rock microparticles (Paterson, 1994; Herwegh and Berger, 2004; Obbard et al., 2011) as these second phases may pin grain boundaries and substantially slow grain growth.

The literature of structural geology is making increasingly use of ideas that come from the material science and information on ice physics (e.g. Weertman, 1983; Petrenko and Whitworth, 1999; Schulson and Duval, 2009). Therefore the present review summarizes first the current understanding of ice deformation and then presents how this understanding may benefit structural geology. Specifically, we concentrate first on the current knowledge of ice characteristics and flow properties, the recrystallization processes and diagnostic microstructures. This is followed by a summary of insights grained into the evolution of microstructure and fabric and their link to processes through 2D in situ experiments. We further provide a summary of recent advances in techniques that have and promise to further enhance the understanding of ice deformation. Here we suggest that synthetic heavy-water (D_2O) ice may be a good analogue for future investigations. The review is concluded by a discussion of how insights from ice have advanced and may in the future enhance our understanding of crustal rock deformation.

THE ANISOTROPIC BEHAVIOUR OF ICE

Only one type of ice (ice 1h) occurs widely in nature on Earth, on a microscopic scale it has a hexagonal close packed (hcp) structure (Pauling, 1935) that is comparable to quartz with a strong anisotropy of basal slip (Fig. 3) and a low Peierls stress (Shearwood and Whitworth, 1991). Both of these minerals deform predominantly by intracrystalline glide on the basal (0001) plane (Steinemann, 1958) and along the ⟨11$\bar{2}$0⟩ direction (Glen and Jones, 1967) and the extent to which this is achieved depend on the orientation of the crystal relative to the main stress axes. An orientation control was proposed by Schmid (1925), with the bulk yield in a single crystal occurring when the resolved shear stress on a slip system reached some critical value (Schmid Law), called the critical resolved shear stress (CRSS). In polycrystalline ice there may be a variation of the yield stress within an aggregate of grains and the validity of the Schmid law (von Mises, 1928) does not hold for all grains in the so called 'easy-slip' orientation for a (0001) type slip (Wilson and Peternell, 2012); as it depends on a grains shape, grain size and grain-boundary networks within a polycrystalline aggregate. It has been found that trace amounts of impurities may produce a marked effect on the CRSS in ice single crystals (Jones and Glen, 1969).

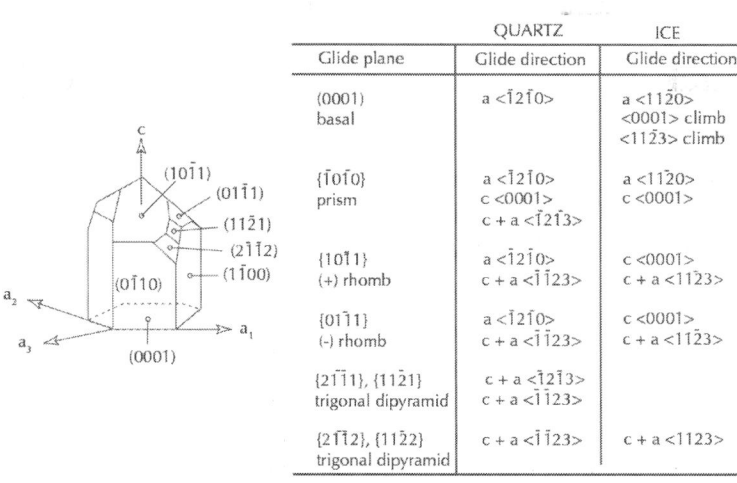

		QUARTZ	ICE
Glide plane		Glide direction	Glide direction
(0001) basal		a <$\bar{1}2\bar{1}0$>	a <11$\bar{2}$0> <0001> climb <11$\bar{2}$3> climb
{$\bar{1}0\bar{1}0$} prism		a <$\bar{1}2\bar{1}0$> c <0001> c + a <$\bar{1}2\bar{1}3$>	a <11$\bar{2}$0> c <0001>
{10$\bar{1}$1} (+) rhomb		a <$\bar{1}2\bar{1}0$> c + a <$\bar{1}$ $\bar{1}$23>	c <0001> c + a <11$\bar{2}$3>
{01$\bar{1}$1} (-) rhomb		a <$\bar{1}2\bar{1}0$> c + a <$\bar{1}$ $\bar{1}$23>	c <0001> c + a <11$\bar{2}$3>
{2$\bar{1}\bar{1}$1}, {11$\bar{2}$1} trigonal dipyramid		c + a <$\bar{1}2\bar{1}3$> c + a <$\bar{1}$ $\bar{1}$23>	
{2$\bar{1}\bar{1}$2}, {11$\bar{2}$2} trigonal dipyramid		c + a <$\bar{1}$ $\bar{1}$23>	c + a <11$\bar{2}$3>

Figure 3: Comparison of slip systems in quartz (after Baëta and Ashbee, 1969) and ice (after Hondoh, 2000). In ice the Burgers vector for the glide motion for edge and screw dislocations is predominantly in (0001) and with climb also in this plane (Hondoh, 2000).

Due to the highly anisotropic behaviour of ice, plastic deformation of polycrystalline ice is significantly influenced by compatibility or accommodation problems between the deformation behaviour of the different individual grains such as: (i) uneven slip between adjacent crystals; (ii) migration of grain boundaries; (iii) kinking of the crystals and (iv) nucleation of new grains. These have significant effects on the development of lattice preferred orientations in polycrystals and vice versa. The Taylor-Bishop-Hill model (Lister et al., 1978) often used to predict crystal orientations is based on free lattice spinning of the crystals in a homogeneously deforming polycrystal and does not require accommodation processes such as dynamic recrystallization. This same situation also applies to the ice fabric simulation models of Van der Veen and Whillans (1994) as they do not involve combined crystallographic slip and dynamic recrystallization.

In this paper we use dynamic recrystallization in the sense described in the geological literature by Urai et al. (1986) when recrystallization is synchronous with deformation. Static recrystallization or annealing is used for grain-boundary rearrangement in the absence of concurrent deformation. It has been suggested that temperature and grain orientation dependence in controlling the yield stress may be explained by the generation of dislocations (Schulson and Duval, 2009). In ice the only dislocations observed to participate in plastic deformation have the

$\langle 11\bar{2}0 \rangle$ Burgers vector lying in the (0001) plane (Shearwood and Whitworth, 1991). Glide is several orders of magnitude easier on basal systems than on non-basal systems (Nakaya, 1958; Wakahama, 1966). The unusual feature of ice is that edge dislocations with the

$\langle 11\bar{2}0 \rangle$ Burgers vector are more mobile on non-basal planes such as $\langle 1\bar{1}01 \rangle$, $\langle \bar{1}101 \rangle$ or $\langle 1\bar{1}02 \rangle$ than on the basal plane (0001) and screw dislocations do not glide on non-basal planes (Hondoh, 2000). At higher temperatures both ice (>13 °C, Wei and Dempsey, 1994) and quartz (>600 °C, Hirth and Tullis, 1992) deform by a

combination of basal and prismatic $\langle 11\bar{2}0 \rangle$ slip and cross-slip of dislocations and this is the key for increasing plasticity in ice

(Chevy et al., 2010). Several theories have been proposed to explain the temperature dependence of yield strength in ice and these are of interest for at least two reasons. The first is that grain-size-sensitive (GSS) flow is controlled by grain-boundary sliding and the second is that grain-size-insensitive (GSI) flow is directly related to the activation of processes involving dislocation motion (Montagnat and Duval, 2000). In rocks De Bresser et al. (2001) suggest a combination of GSI and GSS processes might occur; with GSI recognized by a fabric development, while GSS results in rheological weakening but no fabric development. The ice flow regime in which deformation is accommodated by grain-boundary sliding – GSS (Goldsby and Kohlstedt, 2001; Goldsby, 2006), is generally not recognized in natural ice (Duval and Montagnat, 2002). Instead, grain-size-insensitive dislocation-mediated plasticity is commonly considered as a regular process occurring smoothly in time and homogeneously in space, as suggested by the term "plastic flow" used to describe this plasticity. Below the yield threshold for plastic deformation (Fig. 2) the movement of dislocations comes eventually to a stop, and as a result the material hardens. Above this threshold there is gathering evidence (Chevy et al., 2010) that plastic deformation proceeds through intermittent bursts (avalanches) of activity and recent studies in ice suggest that plasticity is a scale-invariant phenomenon.Chevy et al. (2010) and Montagnat et al. (2006) using X-ray diffraction techniques investigated the dislocation dynamics in ice and found that although such plastic fluctuations may be hard to detect at macro-scales, they dramatically affect mechanical properties at smaller scales, ultimately producing the nuclei for new grains.

FLOW RATE DEPENDENCE OF ICE

One of the most important advances in our understanding of plastic deformation in ice has been the identification, through single crystal and polycrystalline deformation studies, of the exact crystallographic orientation of slip systems (Fig. 3). This has evolved since the first pioneering experimental studies bySteinemann (1958), Barnes et al. (1971) and Kamb (1972) with many hundreds of different experiments that clearly show the relative activity of different slip systems in ice that may be controlled by such variables as temperature and strain rate (Durham et al., 1988; Duval et al., 1983 and references therein). Flow in ice is dominated by the anisotropic slip on the basal-plane (Fig. 1). Therefore in a polycrystalline ice aggregate plasticity must also involve slip or climb of dislocations on non-basal systems. This is analogous to the operation of prism <c> slip in quartz at high homologous temperatures (Hirth and Tullis, 1992), low strain rates and low stress intensity. In ice there is an increased importance of prism <c> slip and diffusion activated processes, that facilitating recovery processes such as grain-boundary migration and dynamic recrystallization to accommodate non-uniform internal stresses within and between grains (Jessell, 1986). The existence of these recovery and recrystallization processes in ice has resulted in many ice laboratory experiments aiming to average the macroscopic behaviour of the ice crystals. This has resulted in the flow-law dependence of polycrystalline ice only being discussed in terms of prescribed stress and strain rates (Glen, 1955; Budd and Jacka, 1989) and has been applied widely in modelling ice dynamics and is often referred to as the 'Glen flow law':

$$\dot{\varepsilon} = B\sigma^n$$

(1)

In the Glen flow law, the stress (σ) exponent n has a value of 3 and B is taken to be constant at a given temperature. Such an analysis is based on a straightforward arithmetic average of the crystal properties to represent aggregate properties to better reproduce realistic characteristics of ice sheets. Many authors use an ad hoc coefficient to proportionately modify Glen›s flow law, which has provided the basis for many numerical descriptions of polycrystalline ice flow (e.g. Budd and Jacka, 1989; Paterson, 1994; Ma et al., 2010). They introduce an empirical coefficient determined for a particular rheological regime, E, called the enhancement factor, this modifies Glen›s flow law as follows:

$$\dot{\varepsilon} = EA\tau_e^{n-1}\tau$$

(2)

which links the strain rate, and deviatoric stress, τ, where A is material dependent parameter, $\tau_e n^{-1} = \tau ij\tau ij/2$ is the square of the second invariant of the deviatoric stress, n is the stress exponent, which depends on the ice temperature via an Arrhenius law. In general, a straightforward average of the crystal properties is not correct, as averaging procedures ignore the local perturbations induced by the grain interaction with neighbours (Tomé, 2000). Similarly, when a fabric develops in an ice aggregate, Glen›s flow law also breaks down as the fabric produces a strong mechanical anisotropy and the strain rate is no longer independent of the orientation of the applied stress. To overcome this Azuma and Higashi (1985) assume glide only occurs on basal systems, and the CRSS on the basal plane of any crystal under uniaxial compression or tension, is related by a geometrical function known as the Schmid Factor, S_g (Azuma and Higashi, 1985).

$$S_g = \cos \phi_0 \sin \phi_0$$

(3)

where ϕ_0 is the angle between the c-axis and the unique stress axis. The Schmid Factor for an aggregate as a whole is given by the mean distribution of c-axes relative to the unique stress axis. Due to the intrinsic anisotropy of ice, the c-axes rotate during deformation towards compressional axes and away from tensional axes (Azuma and Higashi, 1985; Castelnau and Duval, 1994; Van der Veen and Whillans, 1994), and complex feedbacks occur between flow and fabrics (Gagliardini and Meyssonnier, 2000).

Most laboratory ice-deformation experiments have been creep tests performed under constant applied stress, for applied shear stresses lower than 1 MPa and use polycrystalline ice with random c-axis orientations, usually referred to as isotropic ice (e.g. Jacka and Li, 2000; Treverrow et al., 2012). A number of investigators have assumed, without explicit justification, that there is a direct correspondence between the creep curve for constant stress (Fig. 2a) and the stress/strain curve for constant strain rate, although there are some questions about the stress/strain/time histories. However, in the ice physics literature the stress or strain is frequently specified in terms of octahedral shear stress and strain rates (Fig. 2b), where

$$\tau_0 = 1/3 \left[(\sigma_1 - \sigma_2)^2 + (\sigma_2 - \sigma_3)^2 + (\sigma_3 - \sigma_1)^2 \right]^{1/2}$$

(4)

$$\dot{\varepsilon} = 1/3 \left[(\dot{\varepsilon}_1 - \dot{\varepsilon}_2)^2 + (\dot{\varepsilon}_2 - \dot{\varepsilon}_3)^2 + (\dot{\varepsilon}_3 - \dot{\varepsilon}_1)^2 \right]^{1/2}$$

(5)

This may be interpreted in terms of the shear stress across the octahedral shear plane, whose normal has direction cosine $1/\sqrt{3}$ with respect to the principal axes (σ_1, σ_2, σ_3 and ε'_1, ε'_2, ε'_3; Jaeger, 1969). After the initial instantaneous elastic strain the creep rate decreases (Fig. 2b) with time (primary) to a minimum (secondary) at a strain of about 1% (Barnes et al., 1971; Duval et al., 1983; Budd and Jacka, 1989). Beyond the minimum, the creep rate accelerates into a tertiary strain rate (Fig. 2a). The tertiary creep always approaches a steady-state strain rate (Fig. 2b) and the difference between the secondary and tertiary creep stages is identified as the enhancement factor, E, where

$$E = \frac{\dot{\varepsilon}_{ter}}{\dot{\varepsilon}_{min}}$$

(6)

The minimum in strain rate, min, generally occurs at an octahedral shear strain, ϵ_0, of $\approx 1\%$ and is characteristic of the creep deformation of polycrystalline ice (e.g. Wilson and Peternell, 2012), and can be considered the point where microstructural processes controlling recoverable creep and strain hardening are balanced by the onset of recrystallization processes. During primary creep, deformation predominantly occurs by the movement of dislocations on the basal plane (Montagnat et al., 2006), so that the crystal orientation fabric and texture of initially isotropic polycrystalline ice remain unchanged until the strain corresponding to the minimum min (Fig. 2).

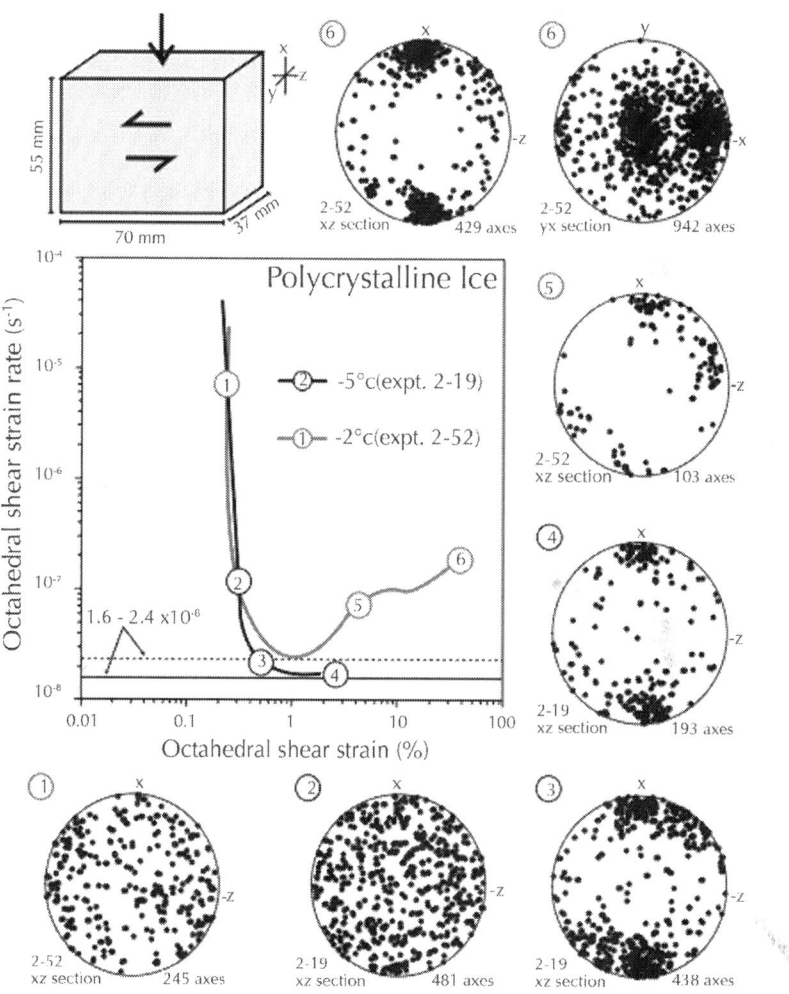

Figure 4: Creep curves and c-axis fabric development for a block of polycrystalline ice deformed at −2 °C and −5 °C that reach minimum shear strain rates of 2.4×10^{-8} and 1.6×10^{-8} s^{-1}, respectively. The ice started with an initial random c-axis fabric and was deformed in combined compression (shortening axis parallel to X) and shear (shear plane YZ) with a compressive stress of 0.22 MPa

and shear stress of 0.4 MPa (after Wilson and Peternell, 2012). The bulk compressive and shear strains for the −2 °C deformation was 10.25% and 0.95 and for the −5 °C experiment was 5.7% and 0.25, respectively. The fabrics at points (1) and (2) record the initial random fabric which remains unchanged during primary creep. Points (3) and (4) display the changes in c-axis fabric change once the minimum or secondary creep is reached at −5 °C. Point (5) represents the fabric transition observed during the accelerating stage of creep. Point (6) is the typical fabric observed during tertiary creep in vertical (XZ section) and horizontal YX sections.

During the transition from secondary (minimum) to tertiary creep dynamic recrystallization processes are activated, resulting in accelerating strain rates and a weakening of the ice (Figs. 4 and 5). At strains of ~10% a dynamic balance between strain hardening and grain-boundary migration and new grain nucleation develops and is associated with the development of distinct crystal orientation fabrics and a steady-state tertiary creep rate. It is now recognized that the geometry of the crystallographic preferred orientation attained in tertiary creep is related to the applied kinematic framework and fabric strengthens with increasing strain. Where dynamic recrystallization is dominant c-axes align symmetrically about the compression direction (Kamb, 1972; Wilson and Russell-Head, 1982). This occurs in both natural ice and in laboratory studies from −20 °C to close to the melting point in the polymorph ice-1h (Petrenko and Whitworth, 1999). In simple shear the fabric maximum is synthetically inclined with the local shear sense and stable at ~70° with respect to the local flow plane. These patterns are comparable to those identified in experimentally deformed quartzites (Tullis et al., 1973; Hirth and Tullis, 1992) and quartz-rich rocks (Kilian et al., 2011; and references therein).

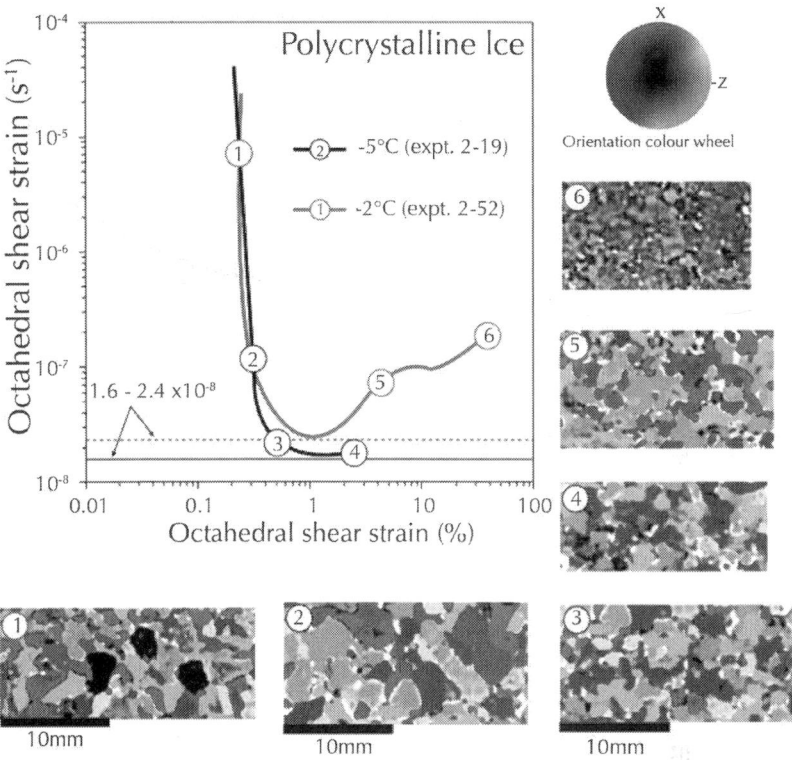

Figure 5: Creep curves and microstructural development in polycrystalline ice deformed at −2 °C and −5 °C. The microstructures are presented as orientation images in XZ sections and colours represent the orientations shown in the colour wheel. The ice started with equiaxed polycrystalline grains that underwent grain-boundary migration during primary creep as observed at points (1) and (2). Points (3) and (4) show the nucleation of discrete new recrystallised grains. Point (5) represents the microstructural transition observed during the accelerating stage of creep, in which there is a combination of grain-boundary migration and new grain nucleation. Point (6) is the typical microstructure observed during tertiary creep consists of an aggregate of polycrystalline grains with a strong c-axis preferred orientation (blue colour).

The transition in microstructural and fabric development in ice from primary through secondary (minimum) to the onset of tertiary creep is clearly seen in the highly anisotropic ice illustrated in Fig. 4, Fig. 5 and Fig. 6. During the phase of primary creep the initial anisotropic grains develop undulose extinction and there is migration of pre-existing grain boundaries (Figs. 5 and 6). At the onset of secondary creep, a new population of randomly oriented recrystallized grains are nucleated at pre-existing grain-boundary margins (Figs. 4 and 6) These new grains are responsible for the destruction of any pre-existing c-axis pattern of preferred orientation and the development of a random pattern (Fig. 6). During the acceleration phase into tertiary creep at strains of ~5–10% there is a dynamic balance between grain-boundary migration and dynamic recrystallization (Fig. 5), and this is associated with the development of distinct crystal orientation fabrics and a steady-state tertiary creep rate. The final crystal orientation fabric that develops is very dependent on the strain regime, in a combined compression and shear deformation there is the development of double maxima fabrics (Fig. 4). On the other hand, unconfined compression experiments are characterized by a clustering of c-axes about a conical surface and form a small-circle girdle pattern about the compression axis. A simple shear regime produces a cluster of a single maximum near-perpendicular to the shear plane. The development of these specific crystal orientation fabrics during tertiary creep, and the pattern illustrated in Fig. 4, are also strongly influenced by initial ice fabrics. Strain rates for anisotropic polycrystalline ice may be up to an order of magnitude higher than corresponding values for isotropic ice deformed under similar conditions (Budd and Jacka, 1989).

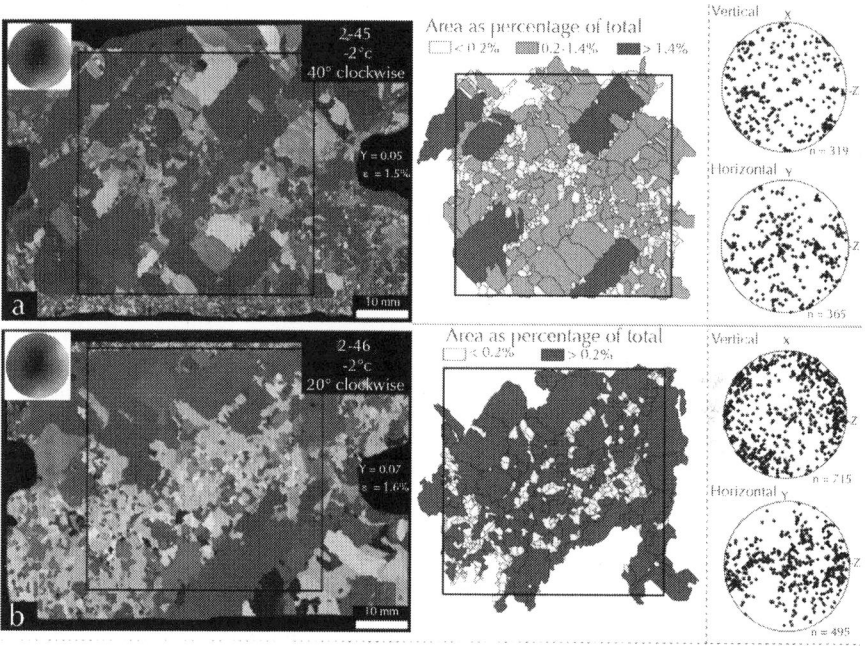

Figure 6: Evolution of microstructure, c-axis fabric and grain statistics from two samples of anisotropic ice deformed −2 °C at small general strains (after Wilson and Peternell, 2012). (a) Orientation image of sample 2-45 where the layering defined by the elongate grains was initially at 40° to the compression direction and has undergone a rotation of 5° with digitized ice crystals (outlined by black rectangle), the orientation colour code relates to the 3D c-axis orientation. Digitized crystals are labelled according to their sizes into three categories of grains are recognized; original elongate grains with undulose extinction (blue), finer original grains with minor migration recrystallization (green) and a population of newly nucleated grains (yellow) that transect the central region of the sample. Stereonets of measured c-axes are equal-area lower hemisphere projections from a vertical and horizontal section through the centre of the deformed sample, the number, n, of measured c-axes is shown at the bottom of the stereonet. (b) Experiment 2-46 the elongate grains were initially at 20° to x_1 and now lie at ~45° to compression axis, and display undulose extinction (green digi-

tized grains) with new grains nucleated at grain boundaries and in a horizontal plane through the centre of the sample (yellow grains in digitized image). Stereonets of measured c-axes are from vertical and horizontal sections through the centre of the deformed sample.

The initial fabric also controls the minimum strain rate (min) with randomly oriented ice having enhancement factors of E ≈ 0, whereas, strongly anisotropic ice have higher enhancement factors E ≈ 5 (Fig. 7). There is also a difference in stress exponents for minimum and tertiary creep regimes that are a function of temperature differences (Fig. 8). Similarly the magnitude of strain-rate enhancement in simple shear has been found to exceed that in uniaxial compression (Fig. 8). Such differences in enhancement factors, which are of great concern to some members of the glaciology community, make little difference to the final tertiary strain rate (Fig. 9), which is dominated by dynamic recrystallization processes that enhance the development of a strong preferred orientation.

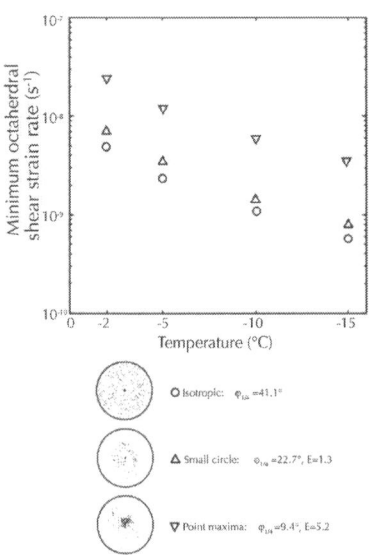

Figure 7: Minimum octahedral shear strain rate (ε̇) for simple shear experiments versus temperature for isotropic polycrystalline ice

and initially anisotropic ice based on data presented by Treverrow et al. (2012). The crystal orientation c -axis data represents the fabric before deformation and $\phi_{1/4}$ is the cone angle containing the first quartile of c -axis colatitudes. Lower values of $\phi_{1/4}$ indicate fabrics that are more strongly vertically clustered. E is the strain-rate enhancement factor (ter/min).

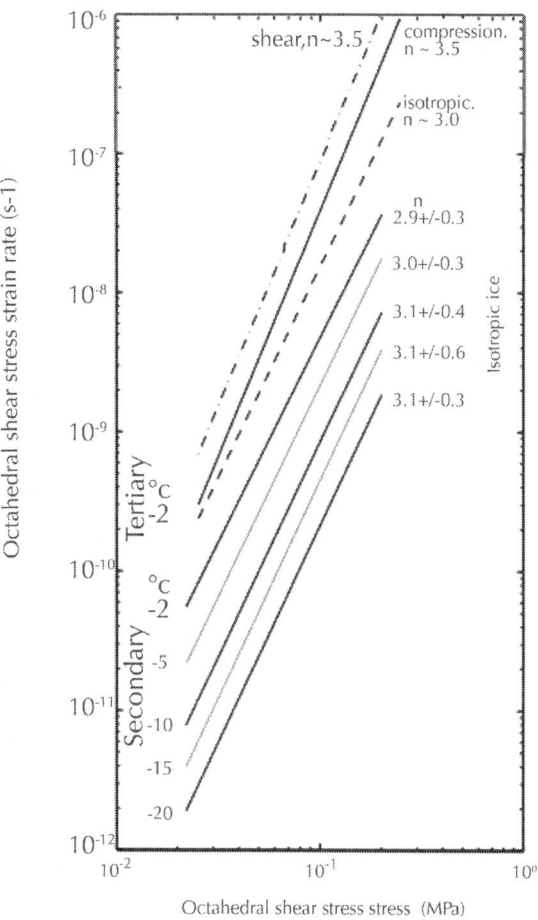

Figure 8: Octahedral shear strain rate $\dot{\varepsilon}$ versus octahedral shear stress (τ_o) for isotropic polycrystalline ice based on data presented by Treverrow et al. (2012) at temperatures from −2 to −20 °C. The

limits on n, the creep power-law stress exponent, for each data set presented by Treverrow et al. (2012) are 95% confidence intervals. The solid lines indicate a least-squares fit to secondary (minimum) octahedral shear strain rate data for each temperature for simple shear experiments that have isotropic minimum creep stress exponents of n = 3.0. The broken lines are tertiary octahedral shear strain data at −2 °C for compression and simple shear experiments with a stress exponent of n = 3.5.

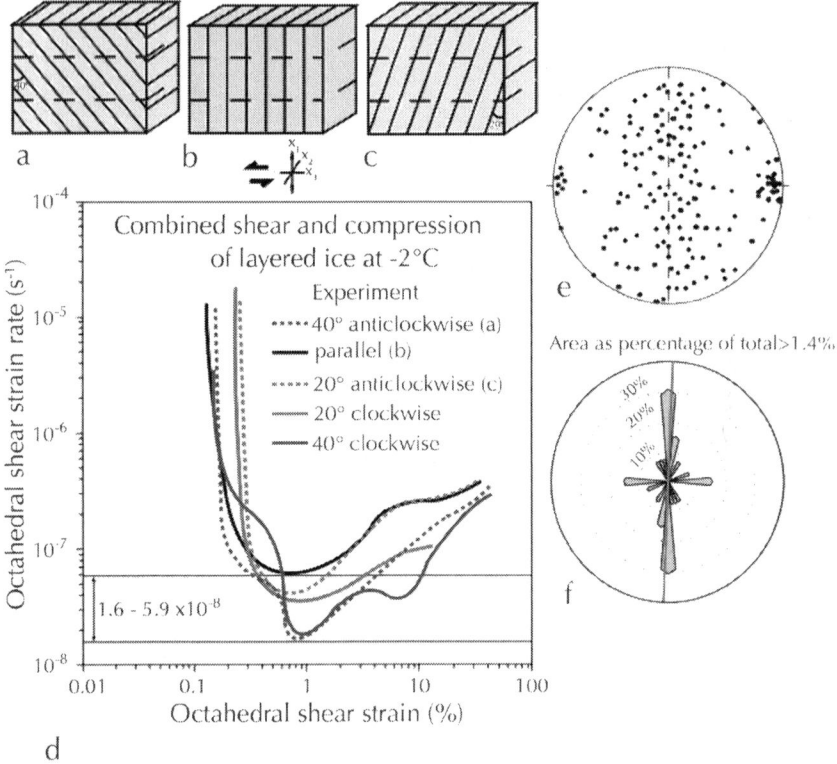

Figure 9: Creep curves for anisotropic ice experimentally deformed at −2 °C in combined compression and shear with a compressive stress of 0.22 MPa and shear stress of 0.4 MPa (a–c) Three examples of initial layered ice configurations under a vertical compressive stress (X$_1$ direction) and anticlockwise shear stress (X$_3$ direction). (d)

The anisotropic ice is dominated by elongate grains that havec-axes lying parallel to the compression axis (2-40), inclined 20° and 40° clockwise (2-46, 2-55), and 20° and 40° anticlockwise (2-43, 2-48) (after Wilson and Peternell, 2012). (e) Example of c-axis fabrics in starting sample (b) with concentration perpendicular to layering. (f) Shape preferred-orientation in (e) as area percentage of grains in each direction.

Fig. 9 is a plot of the yield stresses as a function of orientation for anisotropic ice crystals tested for various orientations of the basal plane. This is analogous to any rock with highly anisotropic grains, and illustrates that a strong variation of CRRS is not only a function of deformation mode but the anisotropy in grain shapes, with elongate grains lying in an 'easy-slip' orientation becoming locked with the consequent hardening of the grains up until the onset of secondary creep (Wilson and Peternell, 2012). After this there is a rapid transition into tertiary creep and the formation of a new crystallographic fabric. The strengthening of this fabric with progressive deformation can result in material weakening through a geometrical softening process (Poirier, 1980), which in ice has been modelled using scalar terms (Azuma and Goto-Azuma, 1996; Thorsteinsson, 2001). This involves a flow law expressed with the strain-rate and stress tensor with a geometrical factor determined by using the c-axis fabric data and stress condition (Azuma and Goto-Azuma, 1996).

A natural creep rate in ice is not only temperature dependent but depends on the magnitude of stress and this also controls the tertiary stress exponent (Goodman et al., 1981). For polar ice dynamics a value of $n = 3.5$ (Fig. 7) for stresses of >1 MPa has been widely used for estimating the flow of naturally deformed ice (Treverrow et al., 2012). There are other laboratory experiments, with stresses of 8–10 MPa and a stress exponent of $n \sim 4$ (Steinemann, 1958; Kirby et al., 1987). Such higher stress data together with grain size analysis have been extrapolated into deformation mechanism maps (Duval et al., 1983), where grain size and flow stress relations generally transect the boundary between the dislocation and diffusion creep fields (Duval et al., 1983). In the case of high-strain rates, involving localized faulting and shearing in ice,Golding et al. (2012) point out that the microstructural state, including the development of dynamic recrystallization during any prior loading history, does not have a significant effect on the microstructural development related to shear localization. Instead they suggest recrystallization probably develops after and is not a cause of shear localization. They suggest it is the critical levels of strain and fast strain rates ($\leq 1 \times 10^{-2}$ s^{-1}) combined with adiabatic heating conditions promote recrystallization. This has implications for some studies of rock and geological materials (e.g. De Bresser et al., 2001) where grain refinement, resulting from recrystallization, is believed to lead to a localized increase in the rate of deformation in the region of refinement, where grain size-sensitive creep (GSS) exceeds that of grain-size-insensitive, power law creep (GSI).

RECRYSTALLIZATION PROCESSES AND DIAGNOSTIC MICROSTRUCTURES

As we have seen dynamic recrystallization involving dislocation mechanisms is an important process that can alter microstructure and rheological behaviour of ice during deformation (Figs. 4 and 5). The basic processes involved in dynamic recrystallization are migration of existing grain boundaries and can lead to grain growth, but can also lead to grain size reduction. This process is in dynamic competition with grain nucleation processes (rotation recrystallization, grain-boundary bulging, grain dissection) that are considered to be important in minerals such as quartz at low homologous temperatures (Stipp et al., 2002). In contrast, at high homologous temperatures (Regime 3, Hirth and Tullis, 1992) grain-boundary migration in quartz is dominant and produces microstructures very comparable to those identified in natural ice masses (Fig. 1a, b). In the ice physics literature not all aspects of the relative role of these processes are clear, but it is generally accepted that dynamic recrystallization is a strain-induced process, no matter whether new grains nucleate or not. For strain-induced recrystallization, without nucleation of new grains, the migrating boundary belongs to an old grain originally present in the polycrystal. This corresponds to the process formerly called 'grain-boundary migration recrystallization' (Beck and Sperry, 1950). This is particularly obvious in the migration of pre-existing grain boundaries where ice is deformed at approximately −20 °C and appears to be the main process to accommodate the basal slip occurring during primary creep (Wilson and Peternell, 2012). In fact, with a plastic strain of ~1% at −20 °C, during primary creep, this is far too small to produce any important rotation of the crystal lattice and subgrain-boundary formation is very low (Hamann et al., 2007). However, once a sample reaches the minimum (~10% strain; Fig. 5), that is secondary creep, there appears to be nucleation of a discrete population of small new recrystalized grains (Fig. 6; yellow grains). In all samples that Wilson and Peternell (2012) examined close to the minimum strain rate there appears to be the nucleation of discrete new grains with a random c-axis preferred orientation (e.g. Fig. 6; yellow grains). Above −15 °C the distinction between grain-boundary migrations versus a discrete grain nucleation stage becomes more difficult (Wilson and Peternell, 2012). During the tertiary creep stage there appears to be further episodes of grain-boundary migration (Fig. 6) that contributes to the final recrystallization history.

There are four different ways in which deformation and dynamic recrystallization are considered to modify microstructure and fabric in an ice aggregate:

- The growth of a polygonized subgrain structure that has achieved misorientation with respect to its surroundings and may be accompanied by undulose extinction (Fig. 1). This can be explained by the formation of a tilt boundary (Wilson, 1986) through the alignment of edge or screw dislocations gliding on the basal plane during bending of the crystal (Hamann et al., 2007; Piazolo et al., 2008). Strain heterogeneities observed inside grains lead to locally high dislocation densities (10^{12}

m^{-2}; Chevy et al., 2010) that accompany the formation of dislocation walls.

- Processes of migration recrystallization (Grain B in Fig. 1c) involve migration through strain-induced grain-boundary migration and the surface energy changes associated with grain-boundary curvature (Montagnat and Duval, 2000). This recrystallization regime is generally referred to as migration recrystallization by geologists and discontinuous recrystallization in materials science. This becomes very important above −5 °C as there is a marked change in activation energy (Barnes et al., 1971), which is associated with large increases in grain-boundary mobility.

- Random recrystallization of new grains with dislocation-free nuclei between deformed grains (Poirier, 1985; Doherty et al., 1997), where the crystallographic orientation of the new grain has no obvious relation to the host grain (Grains D and E in Fig. 1d). Nucleation and/or grain-boundary sliding after nucleation is possible where stored strain (dislocation) energy is high enough. High stored strain energy also provides a driving force, through a relaxation of stored energy, which generates static recrystallization. Strain controls the number of nuclei, and less the rate of growth, so that the recrystallized grain size is primarily dependent on the strain accumulated during deformation (Humphreys and Hatherly, 2004).

- Thermodynamic models for recrystallization of ice under a non-hydrostatic stress field (e.g. Kamb, 1959; MacDonald, 1960). It is assumed that the recrystallized fabric develops by the growth of more favourably oriented crystals at the expense of less favourably oriented and c-axes tend to cluster about a unique stress axis. The manner in which these factors operate to produce a unique recrystallization fabric has been a subject of controversy for more than 50 years and is discussed byPaterson (1973) and Hobbs et al. (2011). Most of the work associated with these models has been superseded by advances in our understanding the development of nuclei involving dislocation networks that originate as small regions in the deformed state (Chevy et al., 2010).

OBSERVATIONS RELEVANT TO UNDERSTANDING MICROSTRUCTURAL DEVELOPMENT IN ICE

Despite the large amount of work that has been done on the mechanical properties of polycrystalline ice, particularly of bulk samples, our understanding of this material on a grain scale has still many serious gaps; particularly in understanding the evolution of microstructure. Another approach has been to deform thin sheets of polycrystalline ice under the optical microscope (Steinemann, 1954; Wakahama, 1964; Burg et al., 1986; Wilson et al., 1986; Wilson and Zhang, 1994). Plane strain 2D in situ deformation experiments of rock analogues and ice allow the direct observation of crystal response to imposed strain, shown by many previous works (e.g. Means, 1980; Urai et al., 1980; Jessell, 1986; Ree, 1991; Ten Brink and Passchier, 1995; Herwegh and Handy, 1996; and many more). In these 2D experiments, the sample has to remain unstrained in the third direction perpendicular to the deformation plane (plane strain) as the thickness of the thin sheet is kept approximately constant by a glass window (Peternell et al., 2011). Because of the strong lattice anisotropy in the ice crystals, parallel to (0001), the initially flat surface becomes roughened by the development of 'valley-ridge structures' that are filled in by a film of silicon oil that is present between the sample and a constraining glass window (Wilson, 1986). These represent the 2D manifestation of glide by basal slip and development is directly related to the crystal orientation relative to the applied stress. The first grains to undergo slip are those oriented for 'easy glide' on the basal plane and can be easily identified by the 'valley-ridge structures' or slip lines (Fig. 10). However, the extent of basal gliding in a given grain is hindered by the geometry of the surrounding crystals and the plane-strain nature of this type of experiment. Therefore the existence of grain boundaries, variations in crystal orientation and response to the applied stress by neighbouring crystals, as a combined effect, suggest that the definition of a favourably oriented crystal is imprecise and inappropriate in this type of experiment. However, this type of experiment does allow continuous and direct observation of a whole range of processes occurring on a microscale.

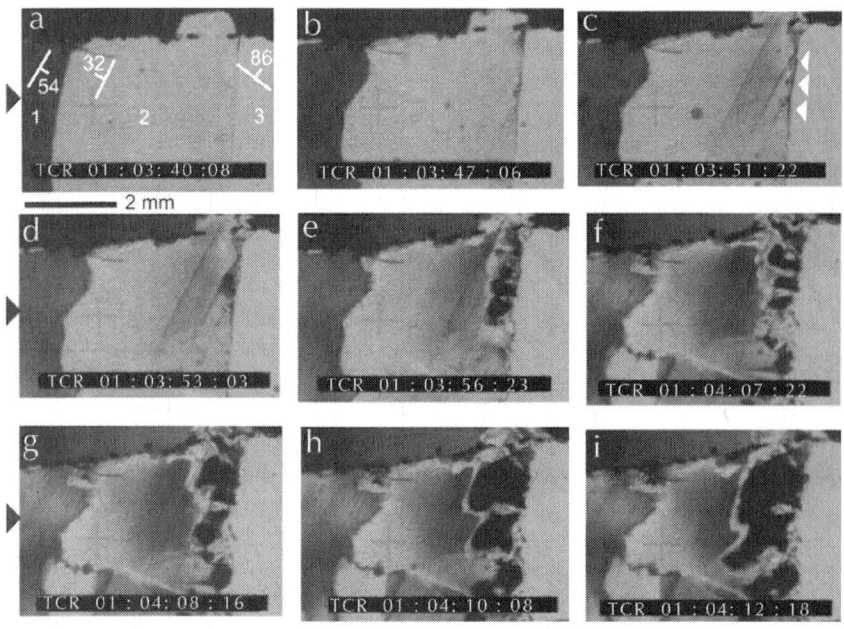

Figure 10: Image sequence from Video 2 showing the ice microstructural evolution in three grains deformed at −5 °C in pure shear. The initial orientation of (0001) in grains 1, 2 and 3 is represented by the dip and strike symbol in (a) and three sites of new grain nucleation are indicated by white arrows in (c). Note the initial grain-boundary migration of grain 1 (a–c), the initiation of slip lines and kinks in grain 2, and the nucleation and growth of new grains (d–i). The arrow on left indicates the shortening direction. The time clock (TCR) reading shows the rapid nucleation and growth of the

new grains. The arrows indicate the shortening direction.

In the section below we will summarize key observations from movies that can be downloaded as part of the supplementary material related to this paper. Three different types of experiments are described: (1) pure and simple shear in situ observations of pure polycrystalline ice, (2) post-deformation annealing of deformed polycrystalline ice, and (3) behaviour of ice with a second phase.

Ice Deformed in Pure Shear

The first obvious change is migration of the original grain boundaries contemporaneous with initiation of slip bands parallel to (0001) (Fig. 10a–c; grain 1). However, intracrystalline deformation within individual crystals is not uniform, as indicated by undulose extinction, and depends on nearest-neighbour interactions. During deformation, the grains are undergoing crystallographic rotation, contemporaneous with the grain-boundary migration. If the adjacent grains restrict slip, the lattice is distorted to produce undulose extinction and kinks. Due to the ongoing deformation (Fig. 10d–f; grain 2), these grains are successively strained and kinks and deformation bands may become particularly well developed in grains that are unsuitably oriented for easy glide.

In naturally deformed ice optically observable kinks are rare, but are commonly developed during 2D in situ experiments. Kinks develop after initial intracrystalline glide, particularly in inequant grains, where basal planes are oriented parallel to the shortening direction. As the deformation, in these 2D experiments, is constrained in the third direction and not homogeneous, such kinks are developed as an experimental artefact. Grains yielding in such a way compensate for shortening that is needed parallel to (0001) in 3D; and the magnitude of lateral migration of the kinks is controlled by the deformation in the neighbouring grains. These kinks are necessary to accommodate any bulk 2D deformation, and are a local strain compatibility feature. However, kinking ceases to be active when kink band boundaries parallel the shortening direction and creation of high-angle boundaries provide sites for new grain nucleation.

Small irregular grains are nucleated adjacent to such grains (e.g. the new grains adjacent to grains 2 and 3 in Fig. 10d–f). New grain nucleation occurs preferentially at intergranular sites, predominantly at high-angle boundaries, and the new grains have orientations significantly different from the adjacent grains. Selective grain growth of these smaller grains into areas of high lattice distortion, presumably higher dislocation density, reduces the area occupied by these unstable grains (Fig. 10g–i). Due to the ongoing deformation, these grains are successively strained and undergo further grain-boundary migration. Such grain boundaries again are the preferred sites for the nucleation of new, dynamically recrystallizing grains. In most in situ experiments, independent of temperature or prior strain rate, a general decaying rate of migration with time is observed decreasing rapidly directly after the deformation.

Dynamic recrystallization observed during in situ experiments is strongly dependent on temperature (seeBurg et al., 1986; Wilson and Zhang, 1996). It is best illustrated at temperatures of −1 °C, where glide on (0001) begins in a favourably oriented grain (Fig. 11a, b), and undulose extinction develops. There is also concurrent nucleation of new grains accompanied by rapid grain-boundary migration (Fig. 11c, d). New grain nucleation occurs preferentially at intercrystalline sites and not generally within grain interiors (intracrystalline sites). However, intracrystalline nucleation sites may occur along kink band boundaries (Fig. 11e); but there is limited evidence that nucleation is preceded by the development of optically distinct subgrains. Due to the ongoing deformation, these new grains are successively strained and the rate of grain-boundary migration decreases before there is a new wave of grain nucleation and boundary migration (Fig. 11e, f). In ice experimentally deformed close to its melting point, for example at ∼ −1 °C, after the cessation of deformation, selected grain boundaries may continue to move through the microstructure in a metadynamic recrystallization stage. That is shortly after deformation stopped (e.g. grain x in Fig. 11f) dynamic recrystallization continues and appears to be independent of stress or temperature increase (Peternell et al., 2013). The onset of such metadynamic grain growth, without a temperature increase, has important implications as to how we interpret the microstructure in ice and rocks, as it appears to be a relaxation of built-up strain energy.

Figure 11: Image sequence from Video 3 showing the microstructural evolution in a polycrystalline grained ice aggregate at −1 °C in pure shear. The arrow on left indicates the shortening direction. (a) Initial microstructure with air bubbles contained in grains 1, 2 and 3. (b) New grain nucleation in grain 1. (c–d) Further new grain nucleation in grain 1 and migration into grain 1 to create two separate grains that subsequently behave as individual entities during further progressive deformation. (e–f) Grain 2 is being consumed by newly nucleated and growing grains. The arrows indicate the shortening direction.

Ice Deformed in Simple Shear

On the scale of a complete sample there may appear to be a homogeneous plastic deformation and this has been described and illustrated by Burg et al. (1986) and Wilson and Zhang (1994). On the scale of individual grains the deformation is very inhomogeneous (Fig. 12a, b) and the evolution of microstructure features are comparable to those recognized in a pure shear deformation. As illustrated in movie 1 and 3 (see supplementary material) and by Burg et al. (1986) grain-boundary migration is the dominant mechanism controlling flow rate at higher temperatures. At lower temperatures (<−10 °C), the processes are too slow to reach a steady-state fabric and microstructure within the duration of the experiment and there are interlocking microstructures with large irregular grains surrounded by smaller new recrystallized grains (Burg et al., 1986).

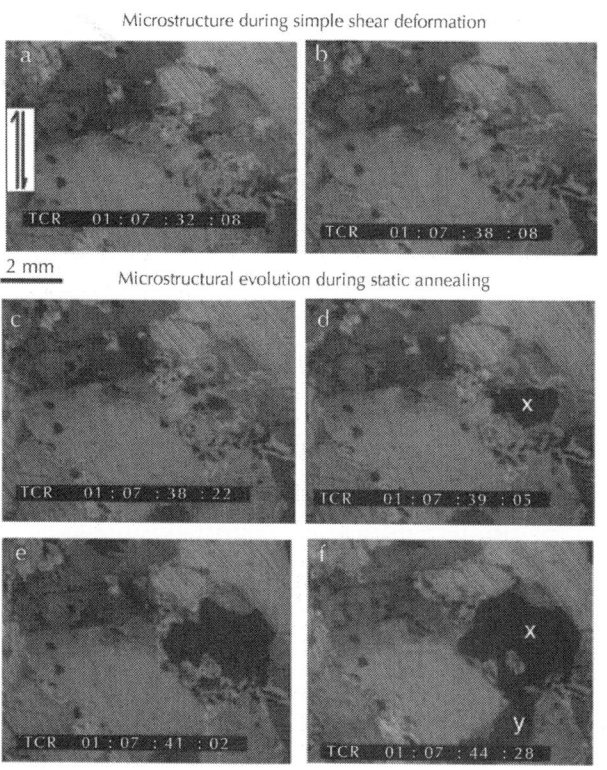

Microstructure during simple shear deformation

Microstructural evolution during static annealing

2 mm

Figure 12: Image sequence of deformation followed by static annealing from Video 6 of a polycrystalline ice deformed in simple shear at −1 °C. The time clock (TCR) reading shows the rapid growth during annealing. (a, b)Microstructures after shear strains of γ = 0.47 and 0.49 and sense of shear is shown in (a). (c − f) Same sample from (b) after annealing in deformation press for 24 h at the same temperature.(f) The grains x and y have annealed forming a polygonal shape and the evolution in the shape change can be traced back to the nuclei in the deformed state (b). The arrows indicate the sense of shear.

Nucleation of small new equiaxed grains appears at very small values of shear strain (< 0.2) with an obvious dependence on temperature (Burg et al., 1986). It is noticeable that the recrystallised grain size at −10 °C is smaller (up to 50%) than that developed in the higher temperature samples (−1 °C). As the shear strain approaches

= 1 continued nucleation and grain-boundary migration can produce a relatively stable shape orientation fabric. As pointed out by Burg et al. (1986) at −4 °C and −1 °C this grain shape fabric might be 16° or 8°, respectively, from the theoretical flattening plane for simple shear. Therefore, estimates of strain magnitude from grain shape and orientation may be unreliable, beyond the fact that the more elongate grains have recorded higher local strains. Hence, in a shear zone defined in part by a continuous reworking of the grain structure by dynamic recrystallization, the significance of grain shape orientation needs to be interpreted with caution.

Post-deformation Annealing of Deformed Ice

Annealing is of utmost importance and is widely recognized in most naturally occurring ice aggregates and often attributed to grain growth occurring as a metadynamic recrystallization stage during a period of lower residual stress (Alley et al., 1986; Rigsby, 1968). It has been shown that in isothermal annealing experiments of deformed ice samples (Wilson, 1982) the time required for annealing to attain a stable grain size decreased with increasing strain and/or temperature (Fig. 13). The annealed grain size attained is directly dependent on the deformed grain size and for ice deformed at temperatures close to its melting point annealing produces insignificant fabric changes (Wilson, 1982). The microstructure of ice deformed at colder temperatures (e.g. −10 °C) and annealed at −1 °C after annealing consisted of polygonal equant grains with smoothly curved or straight interfaces (Wilson, 1982). Although the grain size increases (Fig. 12) the deformation microstructure of ice deformed and annealed at −1 °C is not significantly changed during annealing. Grains are bounded by smoothly curved grain-boundaries and microstructures and crystallographic orientations replicate the grain structure recognized in natural ice (Fig. 1a, b). This suggests that in polar ice sheets and glaciers dynamic recrystallization microstructures may not suffer important modifications while strain rate decays and temperature remains high during static and post-tectonic annealing.

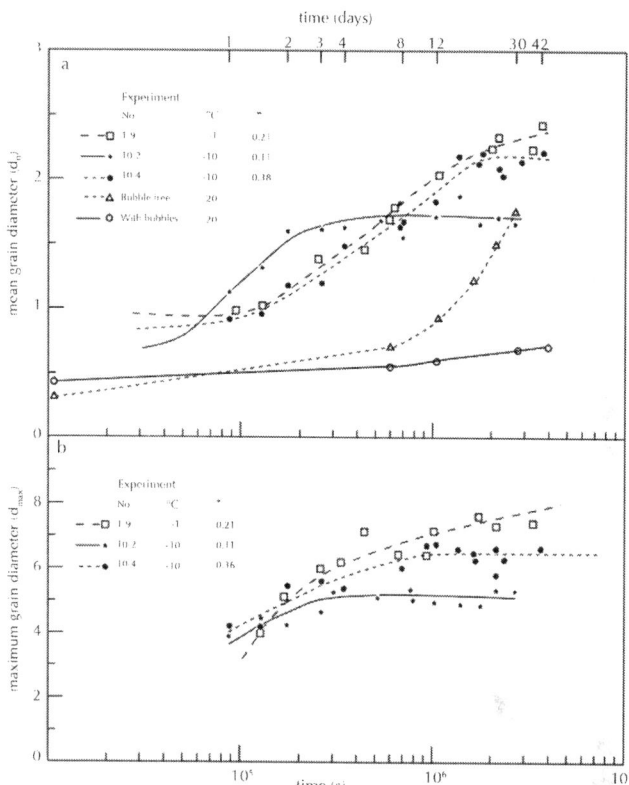

Figure 13: Isothermal annealing curves for polycrystalline ice. (a) Samples deformed at −1 °C and −10 °C and then annealed at −1 °C without a load (after Wilson, 1982), showing mean grain diameter (mm) versus log time for one sample shortened 21% at −1 °C and two samples shortened 11% and 36% at −10 °C. The grain growth curves for bubble free and ice with bubbles at −20 °C (after Azuma et al., 2012) (b) Maximum grain diameter (mm) versus log time for samples described by Wilson (1982).

The rate controlling process in such deformed samples appears to be the build-up of strain, and hence dislocation density and is not surface energy driven. In contrast, in the pure static annealing experiments described by Azuma et al. (2012), grain-boundary diffusion is the rate controlling process. Azuma et al. (2012) demonstrate that slow grain growth rates in ice at −20 °C without bubbles are much greater than with bubbles (Fig. 13a). This influence of a second phase (air bubble) on the grain-boundary migration is called Zener drag, as it was first recognized by Zener (cited in Smith, 1948), and has also been documented by Azuma et al. (2012). The Zener drag relates the volume fraction of second phase to a power law (see Evans et al., 2001 and references therein).

With in situ experiments it is possible to document the nature of static annealing by maintaining or increasing temperature under hydrostatic conditions or at a low residual stress. Post-deformation there may be an initial decrease in grain size attributed to strain energy dissipation as described by Peternell et al. (2013). This is followed by an incubation period, with a nucleation of new grains, which form the nuclei for the development of a new microstructure (Fig. 12c, d). Growth occurs after this incubation period, from pre-existing dynamically recrystallized grains, into the partly or completely recrystallized matrix. The rate of grain-boundary migration decreases with time as these grains grow progressively and consume the host material and presumably the influence of grain-boundary energy increases. This is indicated by decreasing grain-boundary migration rates (Fig. 13), successively straightened grain boundaries, and triple-point junctions that evolve towards equilibrium angles of 120° (Fig. 12e, f).

Deformation of Polycrystalline Ice with a Second Phase

The presence of second phases has major effects and grain growth in ice and generally inhibits the migration of grain boundaries (Obbard et al., 2011) or it can affect the mechanical strength (Jacka et al., 2003). Rheological model experiments (Wilson, 1983) demonstrate that ice grain sizes in an ice-mica mixture were always an order of magnitude smaller than comparable ice-only layers and that grain growth in the ice is pinned by the mica. In such situations the ice occurs as irregular grains bounded by the distribution of the mica platelets, and seldom overgrows the mica. Whereas, the c-axis fabric pattern was identical in the ice-only versus the ice-mica regions (Wilson, 1983); as the deformation was concentrated in the ductile ice matrix, while the enclosed stronger and rigid particles undergo mechanical rotation. Burg and Wilson (1987) have made similar observations using ice-naphthalene mixtures where the degree of intracrystalline slip in the matrix ice is controlled by the distribution and proportion of a second phase. This has been referred to as matrix-controlled deformation (Burg and Wilson, 1987). In matrix-controlled deformation experiments, comparable to the bulk experimental results illustrated in Fig. 10, Jacka et al. (2003) describe polycrystalline ice with sand grain contents up to 15% volume in the stress range of 0.13–0.5 MPa and temperature range −0.02 to −18 °C. They find no significant dependence of either minimum or tertiary flow rate on sand content.

On the other hand where there are weaker objects such as air-bubbles (Azuma et al., 2012) or soft phases, such as camphor (Burg and Wilson, 1987), enclosed in polycrystalline ice they may change shape during the first increments of deformation. In the case of isolated spherical air-bubbles they may not change shape if supported by a rigid ice matrix (e.g. Figs. 11 and 12). The final elongation of the weak object is strongly dependent on the initial orientation of the ice crystal, and the finite stretching direction of the bubbles depends on the amount of slip that is occurring within one individual grain. Grains favourably oriented for glide display more elongated bubbles or enlarge by a coalescence process. Where air-bubbles aggregate or coalesce they tend to elongate in the stretching direction and can be used as strain markers (Hudleston, 1980). With ice-camphor mixtures the weaker camphor-rich domains deform and there is little shape changes in ice microstructure and fabric in ice-rich domains. Such particle-controlled deformationlocalizes in the particle concentrates and the matrix areas show little deformation. Hence the creep dependence on the type of solid impurity and inherent chemical effects with the second phase are highly relevant and are analogous to processes that are observed in rocks (Herwegh and Berger, 2004).

Observations of spherical air-bubbles during an in situ deformation show that their incorporation within grains is dependent on the size and shape of the inclusions and the thickness of the grain boundary (e.g. grains 2 and 3, Fig. 11). During initial grain-boundary migration these inclusions may be left behind (Fig. 11d). Once new grain nucleation and subsequent migration begins they are then incorporated in the new boundary Fig. 11f). As migration is driven by the differences in dislocation density, its rate is very high in the beginning and it sweeps over the inclusion without incorporation. With the nucleation of new grains (Fig. 11e) and during static annealing (Fig. 12f) there is lesser dislocation density, migration is possibly controlled by surface energy differences, and the bubble appears to pin subsequent grain migration as bubbles become incorporated into the boundary.

IN SITU EXPERIMENTS COUPLING DIAGNOSTIC MICROSTRUCTURES AND CRYSTALLOGRAPHIC ORIENTATIONS

In what follows, we include results from in situ deformation undertaken on the stage of an automated Fabric Analyzer (Wilson et al., 2003, 2007) with a deformation press (Peternell et al., 2011). This enabled complete c-axis fabric patterns to be observed continuously and synchronously during an in situ pure shear deformation experiment. The rheological behaviour of the sample was determined at constant strain rate of 2.5×10^{-6} s^{-1}–60% shortening at −10 °C. The sample, RFD_289, was cut from an ice core taken from the Sørsdal Glacier, East Antarctica (Wilson and Peternell, 2011). The area captured during the in situexperiment is characterized by four distinct grains and a cluster of smaller grains that are surrounded by an aggregate of grains (Fig. 14). The microfabric data was recorded from a selected area in the centre of the sample (Fig. 14a) and c-axis orientations can be related to each pixel in the field of view. As a result of Fabric Analyser measurements orientation images, trend flat, geometrical and retardation quality images can be generated (Peternell et al., 2010).

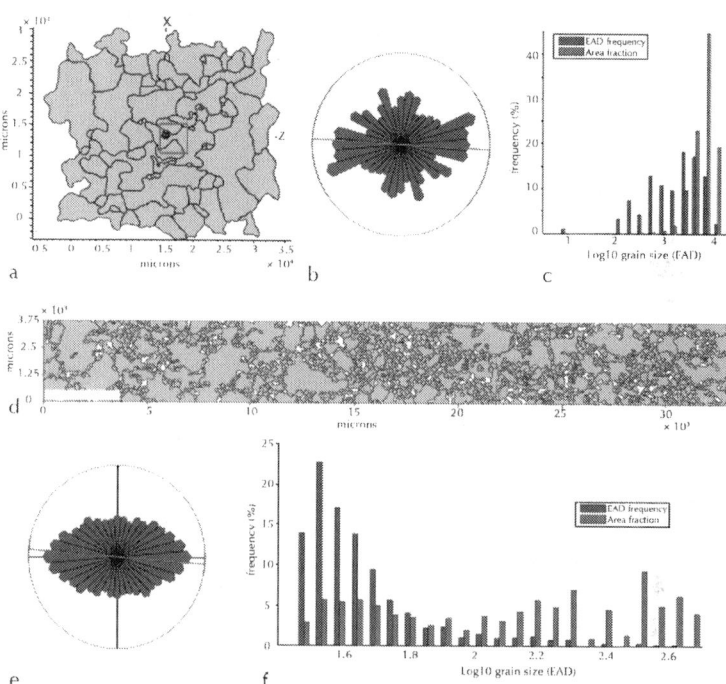

Figure 14: Sørsdal ice at start and completion of in situ experiment RFD_289. (a) Grain shape outline of ice crystals at the start of experiment with area observed during the experiment outlined in red. The shortening direction is X and extension is Z. (b) Shape preferred-orientation analysis of starting material shown as rose diagram the number (n) of measured crystals is 93; red line is mean direction; red circular line is standard deviation. (c) Grain size histogram with frequency distribution of equal-area diameter (EAD; microns) and area fraction occupied by a given EAD class. (d) Manually digitized ice crystals from final stage of deformation. (e) Shape preferred-orientation analysis of the digitized grains shown in rose diagram. (f) Grain size histogram at end of experiment.

The aggregate of grains that made up the sample, initially measured 13.2 × 14.7 mm is deformed in pure shear and the final shape approximates a rectangular shape of 5.1 × 38.1 mm (Fig. 14d). The initial grains were inequant irregular in shape, with no optically visible internal structure. Grain boundaries are gently curved and often intersect at ~120° triple junctions.

Microstructural Evolution

The initial microstructure, in the area of observation, was dominated by four large ice crystals (1–4) accompanied by small crystal aggregates that lie between grains 1 and 2, with individual grains on the boundary between grains 2' and 4 (Fig. 15). The first change in microstructure on straining is the development of serrations in the grain boundaries in the larger grains (grains 1 and 3; Fig. 16a). This begins at ~1% strain and is still obvious at 10% strain (Fig. 16b) and occurs by migration of the original grain boundaries contemporaneous with the development of slip lines parallel to (0001) before there is any marked c-axis orientation change. The first change in the small crystal aggregate (between grains 1 and 2; Fig. 16b) is the migration of pre-existing grain boundaries. Grain size was reduced during grain-boundary migration due to the dissection. For example moving boundaries in grain 3 crosscut or consume parts of neighbouring grain 2' (Fig. 16b).

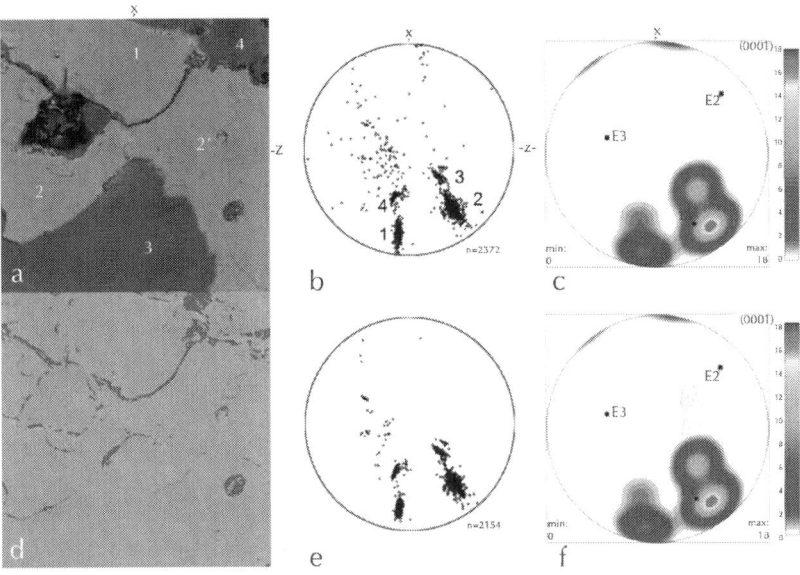

Figure 15: Initial frame from video of ice in situ deformation experiment of natural ice sample RFD_289 and preferred orientation. (a) Starting sample with the 4 initial grains 1 to 4. Image side is 5 mm, X = shortening direction, Z = extension direction. (b) Stereonet of ~2372c-axis orientations selected by a regular grid covering the whole analysed area with concentrations corresponding to 4 initial grains. (c) Contoured orientation distribution. (d) Retardation quality map. (e–f) Stereonet and contoured orientation distribution of 2154 c-axis orientations selected after processing data for geometrical and retardation quality using the methods described byPeternell et al. (2009).

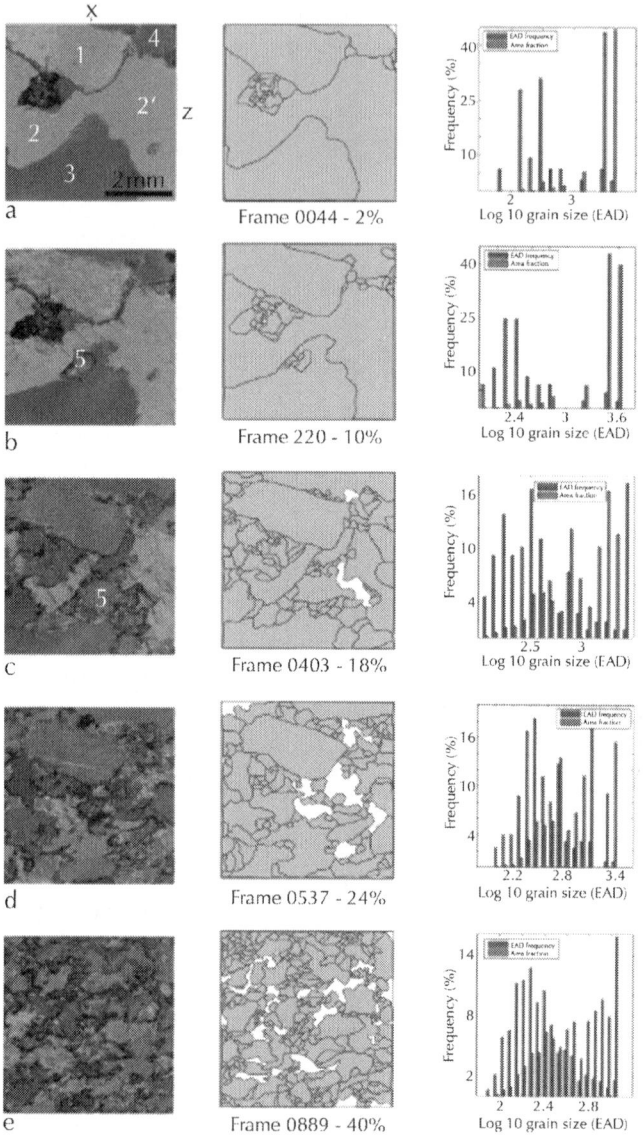

Figure 16: Sequence showing images, grain shape and grain size changes from in situ experiment using natural ice, deformed in pure shear at −10 °C. (a) Grain distribution and shape after larger sample undergoes 2% finite shortening in a vertical direction (×). (b −e) Microstructural evolution at 10, 18, 24 and 40% finite shortening.

Nucleation of small new grains contributes to a grain size reduction but is not a prominent feature at low strain and only becomes obvious at a strain greater than 10% (small grains between 2' and 4 Fig. 16b) and grain 5 (Fig. 16b). Up to a strain of ~18%, the microstructure broadly resembles the starting material. However, the grain 3 begins to develop misoriented subgrains adjacent to boundary with grain 2' (Fig. 16b), these are rapidly consumed by grain migration as grain 5 undergoes initial coarsening (Fig. 16c) before being destroyed by the nucleation of a new grain population (Fig. 16d). The onset of this nucleation stage is what we consider as the beginnings of true dynamic recrystallization.

At a strain of 24% (Fig. 16d) the microstructure is substantially different. The upper half of the sample is dominated by an original grain (grain 1), which displays east-west trending slip lines that lie perpendicular to the compression direction. While the remainder, is dominated by a mixture of small isometric shaped new grains between larger grains with irregular, lobate grain boundaries. A few grains occur throughout some crystals without any obvious relation to any zone of localized deformation. The recrystallization front runs across the entire sample, roughly perpendicular to the compression direction, within this are irregular relics of grains 3 and 4.

At a strain of ~40% (Fig. 16e) the microstructure is completely recrystallized and the distinct recrystallization front has disappeared. The newly nucleated recrystallized grains are generally coarser than at lower strain. Grain boundaries of the larger grains are high-angle, show occasional bulges and vary from wavy to extremely bulged, that suggest grain-boundary migration has again become important. The microstructural evolution with further strain is retained and exhibits the same microstructural features observed at ~60% shortening (Fig. 14 d–f).

The analysis of grain size distributions show that at the start they are very asymmetrical in log-space and continue to be so until 10% strain; during a period dominated by grain-boundary migration (Fig. 16a, b). Once nucleation of new grains occurs the distribution becomes symmetrical (Fig. 16c, d) and the median grain size is reduced and the area fraction of the smaller grains increases (Fig. 16e). Nucleation of new grains by progressive subgrain rotation is of limited importance as evidenced in the lack of subgrain sized grains at grain boundaries. Therefore in these and other high-temperature (>−20 °C) ice experiments (e.g.Wilson, 1986) there is an absence of 'core-mantle structure' often described in natural and experimentally deformed quartzites (e.g. Stipp et al., 2010). The reason for this is that in ice we are dealing with very high homologous temperatures, and need to compare microstructural process during deformation at high metamorphic grades, equivalent to the regime 3 described in the deformation of quartz (Hirth and Tullis, 1992).

Fabric Evolution

Using the Fabric Analyser, the initial c-axis measurements give the orientation of each pixel in the sample (Fig. 15b). Unfortunately, for some pixels the extinction position is not determinable and this leads to ambiguities (rq-retardation quality map). In addition, pixels close to grain boundaries generally give a noisy determination (gq-geometrical quality and rq-retardation quality maps). Therefore by using the filter described by Peternell et al. (2009) the data with gq, rq > 75 can be cleaned up allowing stricter statistical estimates. The c-axis distributions can be presented as the original data set on an equal-area (Schmidt) point plot (Fig. 15b) or as a filtered plot (Fig. 16d) which will always have less c-axes and data can be contoured (Fig. 15c, f). Similarly the eigenvector of the second-order orientation tensor can be defined (Bachmann et al., 2010a) using MTEX (http://code.google.com/p/mtex/) a MATLAB® toolbox (Bachmann et al., 2010b; Mainprice et al., 2011).

The c-axis orientation in the starting material is dominated by three grains, with crystals 1, 2 and 3 represented as three discrete maxima (Fig. 15b, e). With eigenvectors (E_1, E_2, E_3) and maximum contour intervals calculated for each diagram. Fig. 15 shows that after accounting for the gq and rq data the recalculated maxima of 18 and with E_1, E_2 and E_3 vectors plunge 20°–042°, 34°–145°, and 50°–295° respectively. The fabric retains the same pattern with a maximum strength of 15 at 10% shortening (Fig. 17a).With the fabric weakening to a maximum of 10 by 18% shortening (Fig. 17b), through the appearance of randomly oriented c-axes in the centre of the net; together with increased concentrations about the compression axis and development of a four point maxima. Accompanying this is a switch in the position of the E_2 and E_3 vectors (Fig. 17b), with E_2 now plunging 70°–315° and E_3 plunging 5°–070° and E_1 rotating towards the compression axis and becoming more horizontal (18°–164°). However, the E_2eigenvector path is more complicated (Fig. 18a) and corresponds to the episodic nucleation of new recrystallized grains after 10% shortening. The final position of E_2 and E_3 attain some stability between 40 and 60% shortening during which grain-boundary migration dominates.

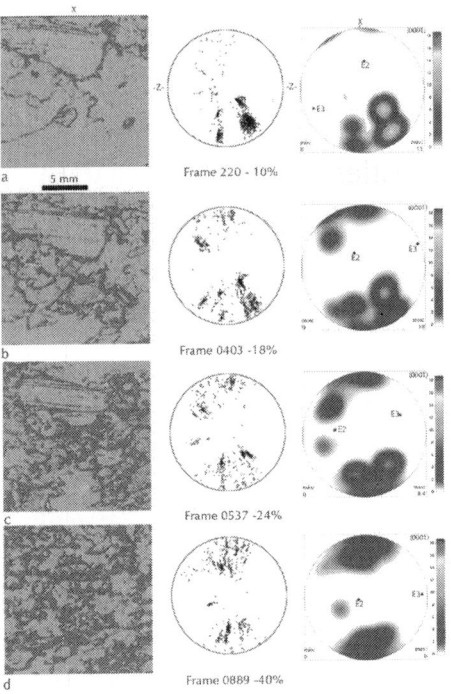

Figure 17: Sequence from in situ experiment using natural ice, deformed in pure shear at −10 °C showing retardation quality im-

ages (rq), stereonet and contoured orientation distribution of c-axis orientations selected after processing data for geometrical and retardation quality using the methods described by Peternell et al. (2009). (a–c) Microstructural and fabric evolution at 10, 18, 24 and 40% finite shortening in a vertical direction (×) and extension (Z).

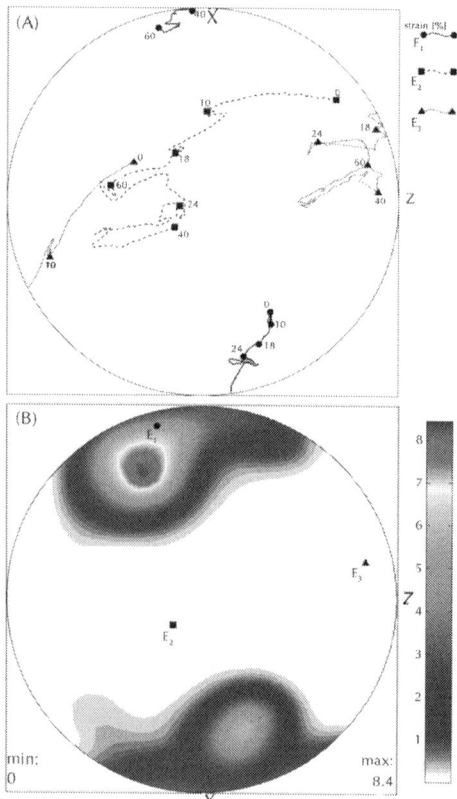

Figure 18: Progressive change in position of eigenvectors and final c-axis fabric for in situ deformation experiment of natural ice sample RFD_289. (A) The change in position of the three eigenvectors E_1, E_2, E_3, calculated from the orientation data, with progressive strain. (B) Final c-axis fabric that covers the whole of sample area (Fig. 15d). This is characterized by two c-axis maxima lying in a small circle about the compression axis (×).

The fabric strength continues to decrease between 18 and 24% shortening (Fig. 17c) with an increased concentration about the compression direction and removal of c-axes that could be related to the original grain 2 (Fig. 15a). By 40% shortening there are fewer c-axes lying outside the concentration that surrounds the compression axis and the distribution is more uniform with the maximum dropping back to 6 while the eigenvectors become into alignment with the compression axis with E_1, E_2 and E_3 vectors plunging approximately 00°–000°, 10°–087°, and 80°–250° respectively (Fig. 18A). The final fabric type after 60% shortening is characterized by two c-axis maxima lying in a small circle about the compression axis with a maximum contoured strength of 5.7 (Fig. 18B); that is comparable to fabrics observed in natural ice (Fig. 1a–b). Such fabric changes to a central c-axis maxima and a strong girdle distribution at ~35° to the compression axis (Fig. 18B) can be attributed to a progressive change in deformation mechanisms from basal slip followed by pyramidal slip.

Deformation and Recrystallization Mechanisms

The microstructures described above show widespread evidence for operation of dislocation processes, such as linear slip lines that is evidence for basal slip in the early stages of deformation. Any subgrain features that may develop are generally masked by the nucleation of new grains. Subgrain structures developed suggest that climb-controlled recovery was also important (e.g. Hamann et al., 2007; Weikusat et al., 2011) probably involving prismatic slip and/or climb. Taking into account the grain size distributions presented in Fig. 16a–b, they show that during the evolution of the microstructure up to 10%, the total strain may be accommodated predominantly by basal slip accompanied by grain-boundary migration, there is little evidence of polygonization process and the formation of subgrains that may be involved in rotation recrystallization (Poirier and Guillopé, 1979.) and may significantly change the overall crystallographic fabric. This is reflected by no drastic change in the c-axis orientation (Fig. 17a, b) up to strains of 10%.

At higher strains it is evident that the fabrics are produced by the operation of a number of slip systems (basal and prismatic) working in conjunction with each other. According to the von Mises condition (von Mises, 1928), there must be deformation by multiple slip systems in order to accommodate the total amount of strain. This, however, can be relaxed with deformation by fracturing, diffusive mass transfer or grain-boundary sliding. In in situ experiments fracturing has been observed (Fig. 9 in Wilson et al., 1986), grain nucleation and boundary migration are pervasive, but clear evidence for grain-boundary sliding remains elusive.

These findings are in agreement with ice deformed at higher temperatures (Burg et al., 1986; Wilson, 1986). Such in situ studies reveal that before the onset of dynamic recrystallization and grain growth there is a similar microstructural evolution to that identified during the bulk deformation of ice (Wilson and Peternell, 2012); namely that during the primary creep phase (<10% strain) there is minimal change in crystallographic preferred orientation. When the deformation proceeds beyond the minimum into tertiary creep regime, the fabric rapidly changes with the onset of new grain nucleation. In the case described here the fabric decreases in intensity (Fig. 17) and the pre-existing crystallographic preferred orientation appears to provide a strong control on the activation of later slip systems. Crystal axes only become aligned with the new kinematic reference frame, that is shortening direction Z, well into the tertiary creep regime (24–40% shortening, Fig. 17d).However, in the case where the initial c-axis fabric was random there is a progressive strengthening (e.g. Burg et al., 1986); as there is a low-angle of rotation of crystallographic axes to accommodate new slip systems during progressive deformation from secondary (~10% shortening, Fig. 17b) into tertiary creep. The in situ observations show that deformation is not homogeneous throughout the polycrystalline aggregate and stress is transmitted in a very complex way and sites of new grain nucleation are almost unpredictable. This means the strain experienced by a particular crystal may differ significantly from the strain observed in the polycrystalline aggregate. It is also important to note that during primary creep dislocation motion in ice is the driving force for grain-boundary migration and that pre-existing preferred orientations may impact on the minimum strain rates at a constant temperature (Fig. 7).

The new grain nucleation observed in ice through in situ experiments is dominantly a dynamic recrystallization phenomenon, by which the stored energy present in a deforming microstructure is used to generate new, dislocation-free grains during the deformation. There appears to be a dynamic balance between recrystallization nucleation rate and grain-boundary migration velocity as described in the model of Derby (1992). There is little optical evidence for rotation recrystallization, which occurs by the gradual increase in subgrain misorientation with deformation strain until subgrains are effectively high-angle grain boundaries (Poirier and Guillopé, 1979). These observations are at odds to a postulation that cannot be verified or observed, namely, rotation recrystallization is a dominant mechanism in many natural ice masses (e.g. Alley, 1992; Thorsteinsson et al., 1997).

The continuous sequence of deformation and recrystallization observed in the in situ experiments gradually results in a steady state, of constant mean grain size (Fig. 16). The activation of new slip systems and grain nucleation is prominent as a strain accommodation process, and there is no evidence for grain-boundary sliding. The suggestion of Goldsby and Kohlstedt (2001) that the deformation of ice at stresses lower than 0.1 MPa with n = 1.8 is dominated by grain-boundary sliding (grain size-sensitive creep) is not in accordance with the observations described above and those described by Montagnat and Duval (2000).

From a range of in situ experiments it can be concluded that dynamic recrystallization by nucleation and grain-boundary migration in ice leads to grain size reduction compared with static conditions and mean grain size increases with increasing temperature (e.g. Weertman, 1983; Jacka and Li, 2000; Wilson and Peternell, 2012). The present results show that rheological weakening in terrestrial ice is best explained by GSI dislocation creep, and is comparable to the hypothesis proposed by De Bresser et al. (1998) where there may be a balance between grain size reduction and grain growth processes set up in the neighbourhood of the boundary between the dislocation and diffusion creep field. The steady-state strain rate and a value of n ≈ 3 (Fig. 8) during dynamic recrystallization are typical for dislocation creep. Grain size reduction by dynamic recrystallization during tertiary creep in either compression or simple shear does not result in a major change in strength or effective viscosity of the ice deforming in steady state (Jacka and Li, 1994). These findings have a major influence on large-scale flow models of ice sheet flow (e.g. Budd and Jacka, 1989: Ma et al., 2010) and influence numerical modelling performed to understand microstructural development.

THE COUPLING OF IN SITU EXPERIMENTS TO NUMERICAL MODELS

Over the last 20 years the traditional barriers that exist between the methodologies of experiments and modelling have been broken down and researchers aim through the integration of in situ experiments and numerical simulations to gain further insight into recrystallization processes (e.g. intracrystalline slip, grain-boundary migration) and their textural and rheological consequences. This scale of modelling is quite different from ice sheet numerical models that are based on the assumption of homogeneous crystallite deformation where the c-axis orientation of each grain can be related (Castelnau and Duval, 1994;Castelnau et al., 1996; Azuma and Goto-Azuma, 1996; Thorsteinsson, 2001). Model microstructures are generally based upon a data structure, that describes the microstructure and crystallographic preferred orientation identified through in situ experiments and relates this to the computed internal stress and dislocation distribution within a grain. Numerical simulations are only as good as the input data.

Classic experimental data provide the input data for numerical simulations in terms of slip system activation; essential data if the crystallographic fabric development in polycrystalline ice aggregates is to be modelled (Van der Veen and Whillans, 1994).

The combination of numerical simulations and in situ experiments is particularly powerful as only in situexperiments provide coherent observations in the form of a time series. These time series are crucial for the validation and extension of numerical simulations, as the post-mortem examination that is available from traditional experimental approaches provide significantly more limited constraints for modelling. The tight coupling of these two types of experiments offers the unique opportunity to test and verify theories for microstructural development, as predictions made by numerical simulations can be directly coupled to appropriate physical experiments, and conversely, theoretical explanations of experimental observations should be testable with numerical simulations (Piazolo et al., 2004). Discrepancies between results obtained with both techniques, suggest the need for an in-depth investigation and thus open up new avenues of theory development, modification and verification.

Once numerical simulations are validated against the in-situ experiments these numerical models can then be use to simulate material behaviours over longer time scales than experimentally possible (Piazolo et al., 2004). Current numerical simulations draw upon recent advances in our ability to numerical simulate subgrain to grain scale processes. Methods include finite-difference methods (Wilson and Zhang, 1994), the Fast Fourier Transformation full field model for crystallographic reorientation (Fig. 19) within a polycrystal (Lebensohn et al., 2007, 2009) with or without a combination of front tracking grain-boundary models (e.g. Jessell et al., 2001; Piazolo et al., 2002; Jessell et al., 2003; Becker et al., 2008).

Wilson and Zhang (1994) used the finite-difference code FLAC, under the assumption that deformation is accommodated by dislocation glide. Models correctly reproduced the deformational and microstructural features caused by glide on (0001) that Wilson and Zhang (1994) observed in situ experiments. In particular triple-point junctions between neighbouring grains were the areas of highest localized stress and strain differences and corresponded to sites where grain-boundary migration was first initiated.

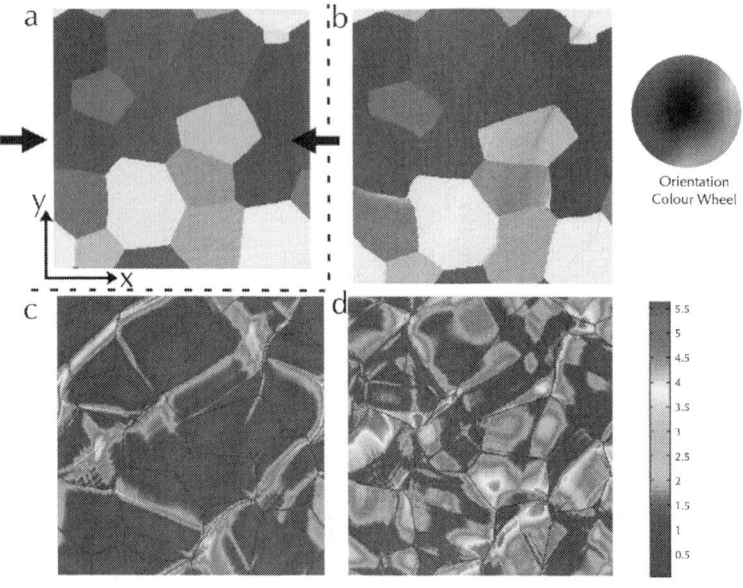

Figure 19: Numerical simulations in polycrystalline ice using the FFT code (after Montagnat et al., 2011; Griera pers comm.). (a) An array of ice grains that are deformed in plane strain with a stress exponent n = 3. (b) Lattice orientation map after 2% shortening. (c) The distribution and magnitude of the strain rate field. (d) The distribution and magnitude of stress field.

More recent studies include detailed work on the prediction of the heterogeneities in orientation changes within a polycrystalline aggregate (Montagnat et al., 2011). Their work shows that experimental results such as continuous kink bands (Fig. 20a), discontinuous subgrain boundaries (Fig. 20b), high variations in strain between adjacent grains can be explained by the highly variable local variation in internal stresses caused by the highly anisotropic behaviour of ice. The model predicts the morphology of the local misorientations within the grains as observed, however the model over-predicts the misorientation values. This is attributed to the absence of modelled annealing and recrystallization mechanisms. A further step was taken by Piazolo et al. (2012) in which numerical simulations take in addition to grain geometries and orientations (a) grain-boundary migration driven by both surface energy and stored energy differences and (b) annihilation of dislocations into account. In this case, results from numerical simulations show a first order coherence with experiments (Figs. 10 and 20). Analysis of numerical experiments show that the discontinuous subgrain boundaries do not originate at grain-boundary asperities, in contrast asperities form due to grain-boundary migration where subgrain boundaries were initiated in response to differences in local dislocation density. Numerical simulations predict slightly higher misorientation across subgrain boundaries than experimentally observed. This can be directly related to the difference in resolution of the numerical model and experimental data (~25 mm versus 1–5 mm).

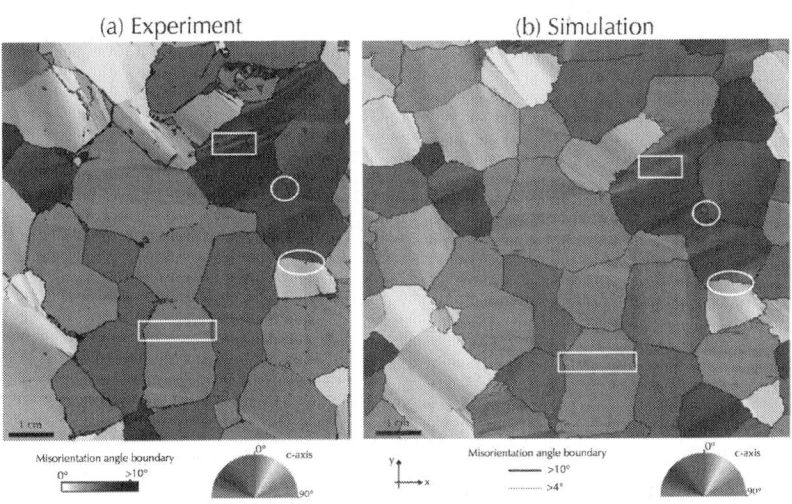

Figure 20: Comparison from ice experiment (a) and numerical simulation (b) after a vertical shortening strain of 4% (after Piazolo et al., 2012). Note that there is a first order correspondence between the results from the experiments to the numerical simulations.

White circles and boxes depict example areas for stress translation i.e. continuation of subgrain boundaries across a grain-boundary and subparallel subgrains crossing whole grains, respectively. Colours indicate the orientation of c-axis; subgrain boundaries >4°, >10°are shown in grey and black.

A NEW PERSPECTIVE FROM EBSD, SYNCHROTRON TOPOGRAPHY AND NEUTRON DIFFRACTION

Measurements of crystallographic fabrics, particularly c-axis fabrics, which were initially confined to versions of an optical microscope and U-stage known as a Rigsby stage (Bader, 1951; Langway, 1958; Rigsby, 1960). This has now reached a considerable level of sophistication employing diverse techniques involving automated optical techniques (Wilen et al., 2003; Wilson et al., 2007; Peternell et al., 2010; Wilson and Peternell, 2011), X-ray, neutron and electron diffraction (Wenk, 2000; Montagnat et al., 2006) as well as electron back scattering diffraction analysis (EBSD; e.g., Iliescu et al., 2004; Piazolo et al., 2008). EBSD analysis enables one to fully characterize the crystallographic texture (both c- and a-axis orientations) of inter- and intracrystalline misorientations in ice (e.g., Iliescu et al., 2004; Obbard et al., 2006; Piazolo et al., 2008; Montagnat et al., 2011; Piazolo et al., 2012; Prior et al., 2012). Such a full crystallographic characterization allows the researcher to deduce possible slip systems and presence of dislocations through trace analysis (e.g., Prior et al., 2002; Barrie et al., 2008; Piazolo et al., 2008).Natural ice shows a spread of 7–8° within individual grains where rotations are dominantly around the c-axis (seeFig. 2c in Piazolo et al., 2008).Analysis shows that observed misorientations can be explained by the activation of either (0001) $\langle 01\overline{1}0 \rangle$ or (0001) $\langle 11\overline{2}0 \rangle$ slip systems with partial edge dislocations in the basal plane with a coupled operation of the two distinct $\langle 11\bar{2}0 \rangle$ Burgers vectors. The latter would require equal activity of both types of dislocations.

Experimentally deformed ice (4% strain at −11 ± 1 °C and 0.5 MPa) exhibits microstructures typically showing straight subgrain boundaries often originating at triple points (Fig. 21a). Using EBSD data these were identified as kink bands with basal edge dislocations in walls that align in contiguous prismatic planes, enabling deformation along the c-axis (Montagnat et al., 2011). In addition, non-uniform grain boundaries (Fig. 21b) discontinuous subgrain boundaries (Fig. 21c) and regions of recrystallization have been observed (Piazolo et al., 2008). Combining EBSD data with the weighted Burgers vector approach, recently proposed by Wheeler et al. (2009), allows a quantitative assessment of dislocation patterns and can be used to reveal the spatial distribution of different types of dislocations within individual grains.

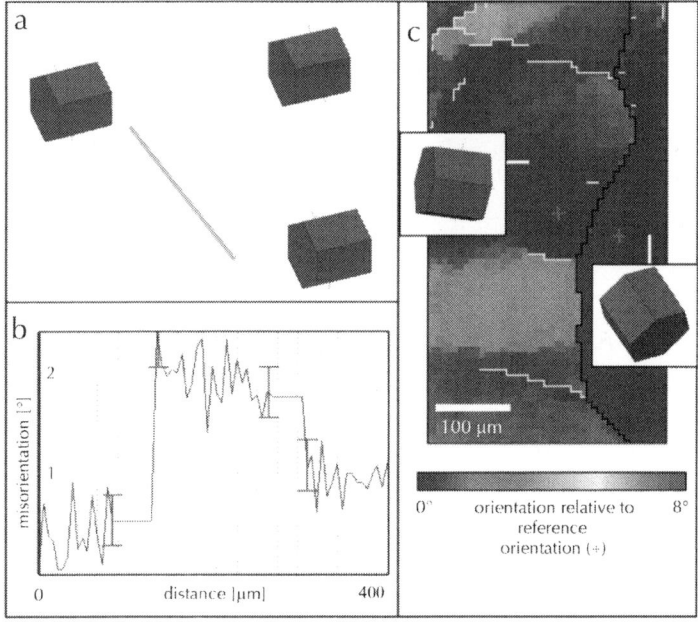

Figure 21: Detailed EBSD analysis from experimentally deformed polycrystalline ice (modified from Piazolo et al., 2012). (a) Sub-

parallel subgrain boundaries as seen in secondary electron image. Subgrain boundaries appear as positive features (ridges) in such images. (b) A transect showing the change in orientation along the grey bar shown in (a) and error bar is ~0.5°. Note that the two subgrains on the upper left and lower right has a similar orientation and are thus subgrain boundaries are identified as kink bands. The 3D crystal orientation as derived through EBSD analysis. (c) Relative orientation map of exemplary grain and subgrain boundaries. Subgrain boundaries of more than 1.5° misorientation (yellow lines) are spatially associated with asperities of grain boundary (black line). Subgrain boundaries are discontinuous and their tip is positioned at curvature changes of the grain boundary. Colour coding is chosen according to relative orientation change with respect to reference orientation (marked with Red Cross).

Ahmad and Whitworth (1988) and Montagnat et al. (2006) have shown that synchrotron X-ray diffraction tomography can be used to investigate the influence of internal stresses on the arrangement of dislocations in ice single crystals during deformation. They point out that for crystals in an easy glide orientation many of the dislocations that define slip lines have a screw character. Cross-slip of these dislocations on non-basal planes induce jog formation and hardening of the ice. This technique allows one to study in detail the bulk characterization of dislocation arrangements and departures from the perfect crystal structure on a resolution of 1 to several microns.

Neutron diffraction analysis which is very useful for full crystallographic characterization of bulk samples (Liang et al., 2009; Covey-Crump et al., 2013) cannot be used for H_2O ice due to its extremely high incoherent scattering. However, it can be conducted on heavy-water (D_2O) where deuterium replaces hydrogen (Bennett et al., 1994). McDaniel et al. (2006) investigated deformation mechanisms in ice using Neutron diffraction analysis on deformed ultra-fine (2–10 µm) grained, polycrystalline heavy-water ice (D_2O). However, results have only a limited application to geological samples due to the extremely small grain size. In contrast, it is possible to use coarse grained (grain size 50–150 μm) polycrystalline D_2O using time-of-flight neutron diffraction during deformation (Covey-Crump and Schofield, 2009). As an ice analogue there is no significant structural difference between D_2O and H_2O (Paterson and Levy, 1957) and both materials exhibit similar rheologies (Fig. 22a). Furthermore fabrication of polymineral samples, for example D_2O ice with calcite, as a two-phase rock is straightforward. First results of experiments with such polymineralic samples show differences in rheological behaviour that can be correlated with microstructural processes (Fig. 22). Even though the stress–strain relationship is steady-state after approximately 10% strain, the presence of calcite extends the process of new grain nucleation (Regime 3;Fig. 22b) in comparison to pure D_2O (Fig. 22a) and is an example of a matrix-controlled deformation (Burg and Wilson, 1987). Furthermore, the application of in situ textural analysis allows one to investigate crystallographic fabric development with progressive finite strain. Fig. 23 reveals interesting fluctuation in textural intensities and character from a random initial c-axis preferred orientation (at 0.25% shortening).At a finite strain of 6.7% shortening the intensity, which correlates with fabric development, increased then decreases at 16.79% shortening.This decrease suggests the nucleation of new recrystallized grains, which progressively grow to produce an increased intensity by 20.5% shortening. The final c-axis pattern of the ice crystals is preferably aligned with the deformation axis (Fig. 23b). The pole figure shows that the (002) poles have changed from random or non-textured orientation to a strong small-circle concentration ofc-axes about the compression axis. A pattern regularly identified in comparable geological (Tullis et al., 1973) and glaciological (Wilson, 1986; Li et al., 1996) experiments.

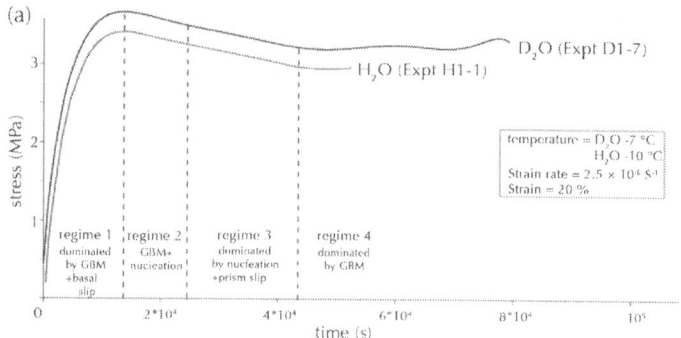

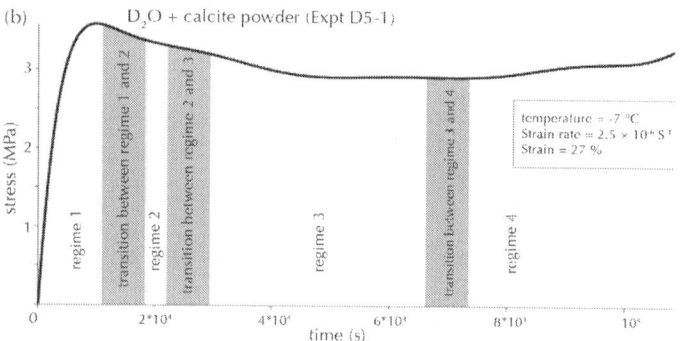

Figure 22: Creep curves for unconfined polycrystalline ice analogues deformed in uniaxial compression. (a) Comparison between pure D_2O ice (Experiment D1-7) and H_2O ice (Experiment H1-1), and (b) D_2O ice and calcite - cc - (Experiment D5-5) all deformed at -7 °C at a strain rate of 2.5×10^{-6} s^{-1}. The microstructural regimes are distinguished on the basis of either dominant grain boundary migration (GBM) or new grain nucleation recognised from textural changes during neutron diffraction analysis and from comparable in situ experiments. For the D_2O ice (a and b), note the initial higher stresses in the two-phase mixture (regime 1 in D5-1) and marked softening at a strain of 2.5–10% (regimes 2 and 3), from 10% strain onwards the strength of the samples remains constant. However, in the two-phase polymineralic sample hardening becomes evident at 27% shortening. Note that even though the stress–strain relationship is steady state after approximately 10% strain, the texture is

changing (cf. Fig. 23a) and this can be attributed to different recrystallization processes (Regimes 1–4).

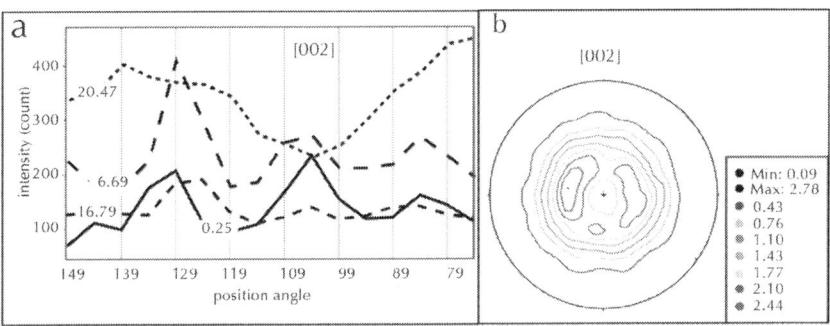

Figure 23: Results from partial texture measurements conducted during a uniaxial deformation experiment of randomly oriented D_2O grains and neutron diffraction analysis. (a) Intensities counted for the [002] axis relative to position angle. Four intensity curves are shown to illustrate the large variability of intensities with progressive deformation. The vertical axis shows intensity scaled by counts, or the number of neutrons reflected off the (002) crystallographic plane at different finite strains. The first intensity measurement was measured at 0.25% shortening and the final intensity was at 20.47% shortening; between these two curves the intensity first increases (6.69% shortening) and then decreases (16.79% shortening) (b) Full pole figure of [002] after deformation showing a distinct small-circle preferred orientation about the compression axis.

USE OF ICE AS ANALOGUE FOR ROCK DEFORMATION

The study of ice deformation has been remarkably successful especially over the last 50 years in enabling the community to understand the creep behaviour at low stresses on materials that have grain sizes and properties similar to many crustal rocks. A fundamental problem in structural geology is to directly link microstructural evolution with rheology. In order to establish this link, experiments are necessary; however, due to the practical aspect of experimental durations, geological materials need to be deformed at high temperatures and pressure at fast strain rates. At the same time, the high temperatures and pressure needed for relevant experiments of geological materials necessitate experiments where we only know the initial and final fabric and microstructural characteristics, but we have no insight into the processes involved during the evolution of the final fabric during progressive deformation unless we undertake a number of successive experiments at varying strain intervals. If we want a clear understanding of these processes then we need to resort to analogue in situ experiments using materials such as ice. At the same time, it is evident that ice is a valuable rock analogue as it replicates microstructural and fabric changes associated with the deformational behaviour of highly anisotropic material with a single dominant slip system as seen for example in quartz or olivine. In the following, we highlight some ways in which ice deformation research helps us to understand the processes ongoing in geological materials such as quartz and olivine. The use of ice can ultimately lead to a system for which various ranges of parameters can be varied and applied that will ultimately produce a range of structures that we are familiar within both naturally and experimentally deformed rocks.

Ice related research has shown that a number of accommodation processes are needed to complement intracrystalline slip in a deforming crystal aggregate. These processes do not co-exist in the same experiment and the orientation of the dominant slip system is critical for the selection of the major accommodation processes. Experiments illustrate the importance of four dominant processes.

- Grain-boundary migration during primary creep. Stress and elastic energy builds up rapidly in crystals and 'badly oriented' grains with higher dislocation densities are then consumed by migration of boundaries. This is a critical process during primary creep and is associated with the consumption of neighbouring grains at boundary regions. It is influenced by the crystallographic setting of the grains relative to each other and by temperature.

- Grain nucleation at the onset of secondary creep. The nature of the nucleation process in ice is attributed to a build-up of dislocation densities and is supported by the recent EBSD results described by Montagnat et al. (2011).

- The existence of high-angle grain boundaries and grain-boundary migration during tertiary creep are the key factors for a ductile accommodation of the imposed deformation. The migration of high-angle boundaries is a ductility-enhancing factor and is achieved by increasing the length of high-angle boundaries. The role of dislocations also accounts for the very high homologous temperatures of the brittle–ductile transition observed in ice and is likely to relate to the very important thermoactivation of the velocity for grain-boundary migration (Urai et al., 1986).

- Post and syn-deformation grain modification is generally considered to be driven by a free energy reduction of the grain boundaries. As pointed out by Roessiger et al. (2012) the microstructure of ice is already strongly affected by dislocation driven processes rather that static grain growth. This observation supports the suggestion by various authors (e.g. Kipfstuhl et al., 2009; Weikusat et al., 2009) that grain-boundary migration and dynamic recrystallization in polar ice precedes static grain growth.

Further insights into rock deformation gained from ice include the mechanisms to accommodate homogeneous strain (von Mises, 1928). In order to accommodate a homogeneous strain that is imposed on aggregate or individual grains it has been clearly shown that in ice single slip on the basal plane operates up to 10% shortening; this implies that the deformation within each grain is inhomogeneous. This in turn leads to the development of heterogeneous recovery and recrystallization processes within a grain in an attempt to match the imposed deformation gradient (Van der Veen and Whillans, 1994). This means that observations from ice aggregates can be extended to the conclusion reached by many structural geologists that slip on a single system is a common process in the development of crystallographic preferred orientations in many rock forming minerals (Schmid, 1994; Reddy and Buchan, 2005; Heilbronner and Tullis, 2006). This is at odds with the postulate made in the Taylor-Bishop-Hill theory of crystallographic preferred orientation development, used by many structural geologists that five independent slip systems operate within each grain (Lister et al., 1978; Lister and Hobbs, 1980.

An important outcome of this review is that it highlights the role of the processes that occur at the transition between the primary and secondary creep regimes. The processes involved during primary creep are similar to those described by Derby and Ashby (1987) where there is an increase of dislocation density, the formation of sub-cells and migration of cell and grain boundaries. This transition exists for a diverse array of processes including, the migration of grain boundaries, the development of crystallographic preferred orientation through the nucleation of a new population of recrystallized grains. What has not been discussed is the effect of a fluid in the system; this raises a whole series of other issues such as what is the role of pressure solution on the integrity of the grain boundary.

In the future, we expect that rock deformation studies will benefit in a number of ways from ice related research. Specifically, the following areas of research can be already identified: (a) the effect of strain rate and thermal history on material properties and microstructures, (b) rheological, microstructural and fabric development in polymineralic rocks and (c) improvement of numerical modelling techniques through ice experiment benchmarking.

In most geological experimental studies strain rates and temperatures may decay rapidly at the termination of an experiment. On the other hand, in both natural rocks and polar ice sheets strain rates may remain high relative to relaxation of thermal histories, so that dynamic recrystallization microstructures may undergo important modifications while strain rate decays and temperature remains high. These complications make it exceptionally difficult to be exact information about the significance of natural microstructures (Bestmann et al., 2005). However, the advantage of using ice analogue experiments is that strain rate and thermal histories in mechanical tests can be better controlled. Similarly the 2D microstructure is particularly good to observe grain interactions with little influence of the third dimension microstructure.

Rheological model experiments using ice and a second phase can be used as analogues for foliation development in quartz-mica rocks where strain is accommodated predominantly by the ice matrix. During particle-controlled deformation, strain is localized in the weaker particle and dissolution through artificial fluid transport mechanisms can create voids (Burg and Wilson, 1987). Such analogue systems can provide a better understanding of plastic deformation and also pressure solution processes that may take place during natural deformation of rocks. The effects associated with phase transformations and chemical reactions involving a second phase, for example clathrate-hydrate systems (Ning et al., 2012), and their interaction with ice has been systematically investigated by Stern et al. (1998) who have observed processes of solid state disproportionation or exsolution under pressure during plastic deformation.

Ice is also a good material to use for improving modelling techniques as it is relatively straightforward to make controlled mechanical tests and simulations of simple microstructures. In particular, the 2D microstructure is particularly well adapted to perform surface observations with little influence of the third dimension. It is clear that a consensus must be reached on the effects in 2D before it is possible to predict them in 3D.

In ice the dominance of grain-boundary migration that accompanies nucleation is in contrast to experiments performed, at lower homologous temperatures, on quartz where three distinct dynamic recrystallization regimes have been identified (Hirth and Tullis, 1992; Stipp et al., 2010); in order of decreasing flow stress, slow (local) grain-boundary migration (regime 1- dislocation creep), subgrain rotation (regime 2), and a combination of boundary subgrain rotation and fast grain-boundary migration (regime 3) occur. A similar sequence of recrystallization microstructures is observed in natural quartz aggregates (e.g. Dunlap et al., 1997; Stipp et al., 2002). As pointed out by Stipp et al. (2010), regimes 2 and 3 are currently not accessible experimentally for quartz, whereas in ice there is a far better understanding of the dynamics of these processes on a submicroscopic scale (e.g. Chevy et al., 2010). This allows us to develop better theoretical models of recrystallization, which incorporate the observed discontinuities in recrystallized grain size and fabric development particularly in regime 3.

We believe that the results obtained from the in situ experiments are relevant to all environments involving polycrystalline aggregates deformed at high temperatures. They are particularly applicable to both geological and glacial environments. Further work in ice deformation and texture analysis includes in situ characterization of texture development at various temperatures and ambient pressure conditions. Similarly the spatial distribution of air-bubbles or inserted strain markers could be used as markers of displacement and hence deformation relative to a chosen point to determine inter-grain strain trajectories. Comparisons can also be made with numerical models; however, adequate account then needs to be taken of other factors, such as the crystallography and the development of anisotropy.

CONCLUSIONS

Ice is an ideal analogue for crustal rocks, as various parameters can be varied and applied during experimental deformation; these will ultimately produce a range of structures that we are familiar to those seen in both naturally and experimentally deformed rocks. Extrapolation of experimental and theoretical implications to natural conditions is sometimes problematic. However, in ice it can be shown from in situ experiments there is a clear switch in strain accommodation mechanisms from grain-boundary migration to new grain nucleation, without polygonization having an important role.

One of the main issues has been whether or not the ice microstructures reflect the most active slip system preserved by the bulk fabric. However the fabrics developed depend not only the operation of a number of different slip systems but the role of competing processes such as grain-boundary migration and new grain nucleation, responsible for microstructural stabilization. It turns out that if we want a clear understanding of these processes then we need to resort to analogue in situ experiments.

ACKNOWLEDGMENTS

This work was supported by the Australian Research Council (Project FT110100070 and DP120102) and the Australian Nuclear Science and Technology Organisation (ANSTO) through ANSTO project 1702 and technical support from the ANSTO staff is gratefully acknowledged. SP acknowledges the support by the Swedish Research Council (VRXX) and the Macquarie University New Staff Grant (9201000756). MicroDICE funding has allowed the authors to interact with their European Science Foundation colleagues during the preparation of this paper. Adam Treverrow, David Prior, Albert Griera and Maurine Montagnat offered helpful comments during the preparation of the paper and supplied versions of Fig. 7, Fig. 8 and Fig. 19 that we subsequently modified. We also thank Marie Diercks for her help in developing the MTEX codes and David Mainprice for providing parts of the MTEX code that allowed us to calculate and plot the eigenvectors. David Prior and Hans De Bresser are thanked for their critical and very helpful reviews.

REFERENCES

1. Ahmad, S., Whitworth, R.W., 1988. Dislocation motion in ice: a study by synchrotron X-ray topography. Philosophical Magazine A 57, 749e766.
2. Alley, R.B., 1992. Flow law hypotheses for ice sheet modelling. Journal of Glaciology 38, 245e256.

3.   Alley, R.B., Perepezko, J.H., Bentley, C.R., 1986. Grain growth in polar ice: I. Theory. Journal of Glaciology 32, 415e424.

4.   Azuma, N., Higashi, A., 1985. Formation processes of ice fabric patterns in ice sheets. Annals Glaciology 6, 130e134.

5.   Azuma, N., Goto-Azuma, K., 1996. An anisotropic flow law for ice-sheet ice and its implications. Annals Glaciology 23, 202e208.

6.   Azuma, N., Miyakoshi, T., Yokoyama, S., Takata, M., 2012. Impeding effects of air bubbles on normal growth of ice. Journal of Structural Geology 42, 184e193.

7.   Bachmann, F., Hielscher, R., Jupp, P.E., Pantleon, W., Schaeben, H., Wegert, E., 2010a. Inferential statistics of electron backscatter diffraction data from within individual crystalline grains. Journal of Applied Crystallography 43, 1338e1355.

8.   Bachmann, F., Hielscher, R., Schaeben, H., 2010b. Texture analysis with mtex e free and open source software toolbox. Solid State Phenomena 160, 63e68.

9.   Bader, H., 1951. Introduction to ice petrofabrics. Journal of Geology 59, 519e536.

10. Baëta, R.D., Ashbee, K.H.G., 1969. Slip systems in quartz: 1. Experiments. American Mineralogist 54, 1551e1593.

11. Barnes, P., Tabor, F.R.S., Walker, J.C.F., 1971. The friction and creep of polycrystalline

12. ice. Proceedings of the Royal Society of London A324, 127e155.

13. Barrie, C.D., Boyle, A.P., Cox, S.F., Prior, D.J., 2008. Slip systems and critical resolved shear stress in pyrite: an electron backscatter diffraction (EBSD) investigation. Mineralogical Magazine 72, 1181e1199.

14. Beck, P.A., Sperry, P.R., 1950. Strain induced grain boundary migration in high purity aluminium. Journal of Applied Physics 21, 150e152.

15. Becker, J.K., Bons, P.D., Jessell, M.W., 2008. A new front-tracking method to model anisotropic grain and phase boundary motion in rocks. Computers and Geosciences 34, 201e212.

16. Bennett, K., Wenk, H.-R., Choi, C.S., Trevino, S.F., Durham, W.B., Stern, L.A., 1994. Texture measurements at 77 K of deformed D2O ice II polycrystals. In:

17. Bunge, H.J., Siegesmund, S., Skrotzki, W., Weber, K. (Eds.), Textures of Geological Materials. DGM Informationsgesellschaft, Uberurel, Germany, pp. 251e275.

18. Bestmann, M., Piazolo, S., Spiers, C.J., Prior, D.J., 2005. Microstructural evolution during initial stages of static recovery and recrystallization: new insights from in-situ heating experiments combined with electron backscatter diffraction analysis. Journal of Structural Geology 27, 447e457.

19. Budd, W.F., Jacka, T.H., 1989. A review of ice rheology for ice sheet modelling. Cold Regions Science and Technology 16, 107e144.

20. Burg, J.P., Wilson, C.J.L., 1987. Deformation of two phase systems with contrasting rheologies. Tectonophysics 135, 199e205.

21. Burg, J.P., Wilson, C.J.L., Mitchell, J.C., 1986. Dynamic recrystallization and fabric development during the simple shear deformation of ice. Journal of Structural Geology 8, 857e870.

22. Castelnau, O., Duval, P., Lebensohn, R.A., Canova, G.R., 1996. Viscoplastic modelling of texture development in polycrystalline ice with a self-consistent approach: comparison with bound estimates. Journal of Geophysical Research 101 (B6), 13,851e13, 868.

23. Castelnau, O., Duval, P., 1994. Simulations of anisotropy and fabric development in Polar ices. Annals Glaciology 20, 277e282.

24. Chevy, J., Fressengeas, C., Lebyodkin, M., Taupin, V., Bastie, P., Duval, P., 2010.

25. Characterizing short-range vs. Long-range spatial correlations in dislocation distributions. Acta Materialia 58, 1837e1849.

26. Covey-Crump, S.J., Schofield, P.F., 2009. Neutron diffraction and the mechanical behaviour of geological materials. In: Liang, L., Rinaldi, R., Schober, H. (Eds.), Neutron Applications in Earth, Energy and Environmental Applications. Springer, New York, pp. 257e282.

27. Covey-Crump, S.J., Schofield, P.F., Stretton, I.C., Daymond, M.R., Knight, K.S., Tant, J., 2013. Monitoring in situ stress/strain behaviour during plastic yielding in polymineralic Rocks using neutron diffraction. Journal of Structural Geology 47, 36e51.

28. De Bresser, J.H.P., Peach, C.J., Reijs, J.P.J., Spiers, C.J., 1998. On dynamic recrystallization during solid-state flow: effects of stress and temperature. Geophysical Research Letters 25, 3457e3460.

29. De Bresser, J.H.P., Ter Heege, T.H., Spiers, C.J., 2001. Grain size reduction by dynamic recrystallization: can it result in rheological weakening? International Journal of Earth Sciences 90, 28e45.

30. Derby, B., 1992. Dynamic recrystallization: the steady state grain size. Scripta Metallurgica et materialia 27, 1581e1586.

31. Derby, B., Ashby, M.F., 1987. A microstructure model for primary creep. Acta Metallurica 35, 1349e1353.

32. Donoghue, S., Jacka, T.H., 2009. The stress pattern within the Law Dome Summit to Cape Folger flow line, inferred from measurements of crystal fabrics. In: Hondoh, T. (Ed.), 2009. Physics of Ice Core Records II, 68. Institute of Low Temperature Sciences, Hokkaido University, pp. 125e135.

33. Doherty, R.D., Hughes, D.A., Humphreys, F.J., Jonas, J.J., Juul Jensen, D., Kassner, M.E., King, W.E., McNelley, T.R., McQueen, H.J., Rollet, A.D., 1997. Current issues in recrystallization: a review. Materials Science and Engineering A238, 219e274.

34. Dunlap, W.J., Hirth, G., Teyssier, C., 1997. Thermomechanical evolution of a ductile duplex. Tectonics 16, 983e1000.

35. Durham, W.B., Kirby, S.H., Heard, H.C., Stern, L.A., Boro, C.O., 1988. Water ice phases II, III, and V: plastic deformation and phase relationships. Journal of Geophysical Research 93, 10191e10208.

36. Duval, P., Ashby, M.F., Anderman, I., 1983. Rate-controlling processes in the creep of polycrystalline ice. Journal Physics and Chemistry 87, 4066e4074.

37. Duval, P., Montagnat, M., 2002. Comments on 'Superplastic deformation of ice: experimental observations' by D.L. Goldsby and D.L. Kohlstedt. Journal of Geophysical Research 107 (B4), 2082, http://dx.doi.org/10.1029/2001JB000946.

38. Evans, B., Renner, J., Hirth, G., 2001. A few remarks on the kinetics of static grain growth in rocks. International Journal of Earth Sciences (Geologische Rundschau) 90, 88e103.

39. Gagliardini, O., Meyssonnier, J., 2000. Simulation of anisotropic ice flow and fabric evolution along the GRIP-GRIP2 flowline, central Greenland. Annals Glaciology 30, 217e223.

40. Glen, J.W., 1955. The creep of polycrystalline ice. Proceedings of the Royal Society of London A228, 519e538.

41. Glen, J.W., Jones, S.J., 1967. The deformation of ice single crystals at low temperatures. In: Oura, H. (Ed.), Physics of Snow and Ice. Institute of Low Temperature Sciences, Hokkaido University, Sapporo, pp. 267e275.

42. Golding, N., Schulson, E.M., Renshaw, C.E., 2010. Shear faulting and localized heating in ice: the influence of confinement. Acta Materialia 58, 5043e5056.

43. Golding, N., Schulson, E.M., Renshaw, C.E., 2012. Shear localization in ice: mechanical response and microstructural evolution of P-faulting. Acta Materially 60, 3616e3631.

44. Goldsby, D.L., 2006. Superplastic flow of ice relevant to glacier and ice-sheet mechanisms. In: Knight, P.G. (Ed.),

Glacier Science and Environmental Change. Blackwell, Oxford, pp. 308e314.

45. Goldsby, D.L., Kohlstedt, D.L., 2001. Superplastic deformation of ice: experimental observations. Journal of Geophysical Research-solid Earth 106 (B6), 11017e 11030.

46. Goodman, D.J., Frost, H.J., Ashby, M.F., 1981. The plasticity of polycrystalline ice. Philosophical Magazine A 43, 665e695.

47. Griera, A., Llorens, M.-G., Gomez-Rivas, E., Bons, P.D., Jessell, M.W., Evans, L.A., Lebensohn, R., 2013. Numerical modelling of porphyroclast and porphyroblast rotation in anisotropic rocks. Tectonophysics 587, 4e29.

48. Hamann, I., Weikusat, C., Azuma, N., Kipfstuhl, S., 2007. Evolution of ice crystal microstructure during creep experiments. Journal of Glaciology 53, 479e489.

49. Hambrey, M.J., 1977. Foliation, minor folds and strain in glacier ice. Tectonophysics 39, 397e416.

50. Heilbronner, R., Tullis, J., 2006. Evolution of c-axis pole figures and grain size during dynamic recrystallization: results from experimentally sheared quartzite. Journal of Geophysical Research 111, B10202.

51. Herwegh, M., Berger, A., 2004. Deformation mechanisms in second-phase affected microstructures and their energy balance. Journal of Structural Geology 26, 1483e1498.

52. Herwegh, M., Handy, M.R., 1996. The evolution of high-temperature mylonitic microfabrics: evidence from simple shearing of a quartz analogue (norcamphor). Journal of Structural Geology 18, 689e710.

53. Hirth, G., Tullis, J., 1992. Dislocation creep regimes in quartz aggregates. Journal of Structural Geology 14, 145e159.

54. Hobbs, B.E., Ord, A., Regenauer-Lieb, K., 2011. The thermodynamics of deformed\ metamorphic rocks: a review. Journal of Structural Geology 33, 758e818.

55. Hondoh, T., 2000. Nature and behavior of dislocations in ice. In: Hondoh, T. (Ed.), Physics of Ice Core Records I. Hokkaido University Press, Sapporo, Japan, pp. 3e24.

56. Hudleston, P.J., 1980. The progressive development of inhomogeneous shear and crystallographic fabric in glacial ice. Journal of Structural Geology 2, 189e196.

57. Humphreys, F.J., Hatherly, M., 2004. Recrystallisation and Related Annealing Phenomena. Elsevier, Oxford, p. 627.

58. Iliescu, D., Baker, I., Chang, H., 2004. Determining the orientations of ice crystals using electron backscatter patterns. Microscopy Research and Techniques 63, 183e187.

59. Jacka, T.H., Li, J., 1994. The steady-state crystal size of deforming ice. Annals Glaciology 20, 13e18.

60. Jacka, T.H., Li, J., 2000. Flow rates and crystal orientation fabrics in compression of polycrystalline ice at low temperature and stresses. In: Hondoh, T. (Ed.), Physics of Ice Core Records I. Hokkaido University Press, Sapporo, Japan, pp. 83e113.

61. Jacka, T.H., Donoghue, S., Li, J., Budd, W.F., Anderson, R.M., 2003. Laboratory studies of the flow rates of debris-laden ice. Annals Glaciology 37, 108e112.

62. Jaeger, J.C., 1969. Elasticity, Fracture and Flow with Engineering and Geological Applications. (Methuen, London).

63. Jessell, M.W., 1986. Grain boundary migration and fabric development in experimentally deformed octachloropropane. Journal of Structural Geology 8, 527e542.

64. Jessell, M.W., Bons, P., Evans, L., Barr, T., Stuwe, K., 2001. Elle: the numerical simulation of metamorphic and deformation microstructures. Computers and Geosciences 27, 17e30.

65. Jessell, M.W., Kostenko, O., Jamtveit, B., 2003. The preservation potential of microstructures during static grain growth. Journal of Metamorphic Geology 21, 481e491.

66. Jones, S.J., Glen, J.W., 1969. The effects of dissolved impurities on the mechanical properties of ice crystals. Philosophical Magazine 19, 13e24.

67. Kamb, W.B., 1959. Theory of preferred crystal orientation developed by crystallization under stress. Journal of Geology 67, 153e170.

68.   Kamb, W.B., 1972. Experimental recrystallization of ice under stress. In: Heard, H.C., Borg, I.Y., Carter, N.L., Raleigh, C.B. (Eds.), Flow and Fracture of Rocks, Geophysical Monograph, 16. American Geophysical Union, Washington D.C,

69.   pp. 211e214.

70.   Kilian, R., Heilbronner, R., Stünitz, H., 2011. Quartz microstructures and crystallographic preferred orientation: which shear sense do they indicate? Journal of Structural Geology 33, 1446e1466.

71.   Kirby, S.H., Durham, W.B., Beeman, M.L., Heard, H.C., Daley, M.A., 1987. Inelastic properties of ice 1h at low temperatures and high pressures. Journal of Physics 48, 227e232.

72.   Kipfstuhl, S., Faria, S.H., Azuma, N., Freitag, J., Hamann, I., Kaufmann, P., Miller, H., Weiler, K., Wilhelms, F., 2009. Evidence of dynamic recrystallization in polar firn. Journal of Geophysical Research 114 (B5), B05204. http://dx.doi. org/10: 1029/2008JB005583.

73.   Liang, L., Rinaldi, R., Schober, H., 2009. Neutron Applications in Earth, Energy and Environmental Applications. Springer, New York.

74.   Langway, C.C., 1958. Ice Fabrics and the Universal Stage, p. 16. CREEL Technical Report 62.

75.   Lebensohn, R.A., Tomé, C.N., Ponte Castañeda, P., 2007. Self-consistent modelling of the mechanical behaviour of viscoplastic polycrystals incorporating intragranular field fluctuations. Philosophical Magazine 87, 4287e4322.

76.   Lebensohn, R.A., Montagnat, M., Mansuy, P., Duval, P., Meysonnier, J., Philip, A., 2009. Modeling viscoplastic behavior and heterogenous intracrystalline deformation of columnar ice polycrystals. Acta Materialia 57, 1405e1415.

77.   Li, J., Jacka, T.H., Budd, W.F., 1996. Deformation rates in combined compression and shear for ice which is initially isotropic and after the development of strong anisotropy. Annals of Glaciology 23, 247e252.

78. Lister, G.S., Paterson, M.S., Hobbs, B.E., 1978. The simulations of fabric development in plastic deformation and its application to quartzite: the model. Tectonophysics 45, 107e158.

79. Lister, G.S., Hobbs, B.E., 1980. The simulation of fabric development during plasticdeformation and its application to quartzite e the influence of deformation history. Journal of Structural Geology 2, 355e370.

80. Ma, Y., Gagliardini, O., Ritz, C., Gillet-Chaulet, F., Durand, G., Montagnat, M., 2010. Enhancement factors for grounded ice and ice shelves inferred from an anisotropic ice-flow model. Journal of Glaciology 56, 805e812.

81. MacDonald, G.J.F., 1960. Orientation of anisotropic minerals in a stressfield. In: Griggs, D.T., Handin, J.W. (Eds.), Rock Deformation, Geological Society of America Memior, 79, pp. 1e8. Bolder, Colorado.

82. Mainprice, D., Hielscher, R., Schaeben, H., 2011. Calculating anisotropic physical properties from texture data using the MTEX open source package. In: Prior, D.J., Rutter, E.H., Tatham, D.J. (Eds.), Deformation Mechanisms, Rheology and Tectonics: Microstructures, Mechanics and Anisotropy, Geological Society, London, Special Publications, 360, pp. 175e192.

83. Marmo, B.A., Wilson, C.J.L., 1998. Strain localisation and incremental deformation within ice masses, Framnes Mountains, east Antarctica. Journal of Structural Geology 20, 149e162.

84. McDaniel, S., Bennett, K., Durham, W.B., Waddington, E.D., 2006. In situ deformation apparatus for time-of-flight neutron diffraction: texture development of polycrystalline ice Ih. Review of Scientific Instruments 77, 093902, 1e6.

85. Means, W.D., 1980. High-temperature simple shearing fabrics e a new experimental approach. Journal of Structural Geology 2, 197e202.

86. Montagnat, M., Duval, P., 2000. Rate controlling processes in the creep of polar ice, influence of grain boundary migration associated with recrystallization. Earth Planet. Science Letters 183, 179e186.

87. Montagnat, M., Weiss, J., Chevy, J., Duval, P., Brunjail, H., Bastie, P., Sevillanos, J.G., 2006. The heterogeneous nature of slip in ice single crystals deformed under torsion. Philosophical Magazine 86, 4259e4270.

88. Montagnat, M., Blackford, J.R., Piazolo, S., Arnaud, L., Lebensohn, R.A., 2011. Measurements and full-field predictions of deformation heterogeneities in ice. Earth and Planetary Science Letters 305, 153e160.

89. Nakaya, U., 1958. The Deformation of Single Crystals of Ice, 47. IAHS Publication, pp. 229e240.

90. Ning, F., Yu, Y., Kjelstrup, S., Vlugt, T.J.H., Glavatskiy, K., 2012. Mechanical properties of clathrate hydrates: status and perspectives. Energy Environmental Science 5, 6779e6795.

91. Nye, J.F., 1953. The flow law of ice from measurements in glacier tunnels, laboratory experiments and thr Jungfraufirn borehole experiment. Proceedings of the Royal Society of London A219, 477e489.

92. Obbard, R.W., Baker, I., Sieg, K., 2006. Using electron backscatter diffraction patterns to examine recrystallization in polar ice sheets. Journal of Glaciology 52, 546e557.

93. Obbard, R.W., Baker, I., Prior, D.J., 2011. A scanning electron microscope technique for identifying the mineralogy of dust in ice cores. Journal of Glaciology 57, 511e514.

94. Paterson, S.W., Levy, H.A., 1957. A single-crystal neutron diffraction study of heavy ice. Acta Crystalographica 10, 70e76.

95. Paterson, M.S., 1973. Thermodynamics and its geological application. Reviews of Geophysics and Space Physics 11, 355e389.

96. Paterson, W.S.B., 1994. The Physics of Glaciers, third ed. Pergamon, Oxford, p. 480.

97. Pauling, L., 1935. The structure and entropy of ice and of other crystals with some randomness of atomic arrangement. Journal of American Chemical Society 57, 2680e2684.

98. Peternell, M., Kohlmann, F., Wilson, C.J.L., Seiler, C., Gleadow, A.J.W., 2009. A new approach to crystallographic orientation measurement for apatite fission track analysis: effects of crystal morphology and implications for automation. Chemical Geology 265, 527e539.

99. Peternell, M., Hasalová, P., Wilson, C.J.L., Piazolo, S., Schulmann, K., 2010. Evaluating quartz crystallographic preferred orientations and the role of deformation partitioning using EBSD and fabric analyser techniques. Journal of Structural Geology 32, 803e817.

100. Peternell, M., Russell-Head, D.S., Wilson, C.J.L., 2011. A technique for recording polycrystalline structure and orientation during in situ deformation cycles of rock analogues using an automated Fabric Analyser. Journal of Microscopy 242, 181e188.

101. Peternell, M., Diercks, M., Wilson, C.J.L., Piazolo, S., 2013. Quantification of polycrystalline fabric, using FAME: application to in situ deformation of ice. Journal of Structural Geology.

102. Petrenko, V.F., Whitworth, R.W., 1999. Physics of Ice. Oxford Univ. Press, Oxford.

103. Piazolo, S., Bons, P.D., Jessell, M.W., Evans, L., Passchier, C.W., 2002. Dominance of Microstructural Processes and Their Effect on Microstructural Development: Insights from Numerical Modelling of Dynamic Recrystallization. In: Geological Society of London, Special Publications, 200, pp. 149e170.

104. Piazolo, S., Jessell, M.W., Prior, D.J., Bons, P.D., 2004. The integration of experimental in-situ EBSD observations and numerical simulations: a novel technique of microstructural process analysis. Journal of Microscopy 213, 273e284.

105. Piazolo, S., Montagnat, M., Blackford, J.R., 2008. Substructure characterization of experimentally and naturally deformed ice using cryo-EBSD. Journal of Microscopy 230, 509e519.

106. Piazolo, S., Borthwick, V.E., Griera, A., Montagnat, M., Jessell, M.W., Lebensohn, R., Evans, L., 2012. Substructure dynamics in crystalline materials: new insight from in-situ experiments, detailed EBSD analysis of experimental and natural samples and numerical modelling. Material Science Forum 715-716, 502e507.

107. Poirier, J.P., 1980. Shear localization and shear instability in materials in the ductile field. Journal of Structural Geology 2, 135e142.

108. Poirier, J.P., 1985. Creep of Crystals. Cambridge University Press, Cambridge, p. 260.

109. Poirier, J.P., Guillopé, M., 1979. Deformation-induced recrystallization of minerals.

110. Bulletin Mineralogie 102, 67e74.

111. Prior, D.J., Wheeler, J., Peruzzo, L., Spiess, R., Storey, C., 2002. Some garnet microstructures: an illustration of the potential of orientation maps and misorientation analysis in microstructural studies. Journal of Structural Geology 24, 999e1011.

112. Prior, D.J., Diebold, S., Obbard, R., Daghlian, C., Goldsby, D.L., Durham, W.B., Baker, I., 2012. Insight into the phase transformations between ice Ih and ice II from electron backscatter diffraction data. Scripta Materialia 66, 69e72.

113. Ree, J.H., 1991. An experimental steady-state foliation. Journal of Structural Geology 13, 1001e1011.

114. Reddy, S.M., Buchan, C., 2005. Constraining kinematic rotation axes in high-strain zones: a potential microstructural method? In: Gapais, D., Brun, J.P., Cobbold, P.R. (Eds.), Deformation Mechanisms and Tectonics from Minerals to the Lithosphere, Geological Society of London, Special Publications, vol. 243, pp.1e10.

115. Rigsby, G.P., 1960. Crystal orientation in glacier and in experimentally deformed ice. Journal of Glaciology 3, 589e606.

116. Rigsby, G.P., 1968. The complexities of the three-dimensional shape of individual crystals in glacier ice. Journal of Glaciology 7, 233e252.

117. Roessiger, J., Bons, P.D., Griera, A., Jessell, M.W., Evans, L., Montagnat, M., Kipfstuhl, S., Faria, S.H., Weikusat, I., 2012. Competition between grain growth and grain size reduction in polar ice. Journal of Glaciology 57, 942e948.

118. Schmid, S.M., 1994. Textures of geological materials: computer model predictions versus empirical interpretations based on rock deformation experiments and field studies. In: Bunge, H.J., Siegesmund, S., Skrotzki, W., Weber, K. (Eds.), Textures of Geological Materials. DGM Informationsgesellschaft, Uberurel, Germany, pp. 279e301.

119. Schmid, E., 1925. Zn-normal stress law. Proceedings International Congress of

120. Applied Mechanics, Delft 1924, p.342.

121. Schulson, E.M., Duval, P., 2009. Creep and Fracture of Ice. Cambridge University Press, Cambridge.

122. Shearwood, C., Whitworth, R.W., 1991. The velocity of dislocations in ice. Philosophical Magazine A64, 289e302.

123. Smith, C.S., 1948. Grains, phases, and interphases: an interpretation of microstructure. Transactions American Institute of Mining and Metallurgical Engineering 175, 15e51.

124. Steinemann, S., 1954. Results of preliminary experiments on the plasticity of ice crystals. Journal of Glaciology 2, 404e413.

125. Steinemann, S., 1958. Experimentalle Untersuchungen zur Plastizität von Eis. Beitr. Geol.. der Schweiz, geotech. Ser., Hydrol 10, 1e72.

126. Stern, L.A., Kirby, S.H., Durham, W.B., 1998. Polycrystalline methane hydrate: synthesis from superheated ice, and low-temperature mechanical properties. Energy and Fuels 12, 201e211.

127. Stipp, M., Stunitz, H., Heilbronner, R., Schmid, S.M., 2002. Dynamic recrystallization of quartz: Correlation between natural and experimental conditions. In: De Meer, S., Drury, M.R., De Bresser, J.H.P., Pennock, G.M. (Eds.), Deformation Mechanisms, Rheology and Tectonics: Current Status and Future Perspectives, Geological Society Special Publication London, v. 200, pp. 171e190.

128. Stipp, M., Tullis, J., Scherwath, M., Behrmann, J.H., 2010. A new perspective on paleopiezometry: dynamically recrystallized grain sizes distributions indicate mechanism changes. Geology 38, 759e762.

129. Taupin, V., Richeton, T., Chevy, J., Fressengeas, C., Weiss, J., Louchet, F., Miguel, M.C., 2008. Rearrangement of dislocation structures in the aging of ice single crystals. Acta Materialia 56, 1555e1563.

130. Ten Brink, C.E., Passchier, C.W., 1995. Modelling of mantled porphyroclasts using non-Newtonian rock analogue materials. Journal of Structural Geology 17, 131e146.

131. Thorsteinsson, T., 2001. An analytical approach to deformation of anisotropic icecrystal aggregates. Journal of Glaciology 47, 507e516.

132. Thorsteinsson, T., Kipstuhl, J., Miller, H., 1997. Textures and fabrics in the GRIP ice core. Journal of Geophysical Research 102 (C12), 26,583e26, 599.

133. Tomé, C.N., 2000. Tensor properties of textured polycrystals. In: Kocks, U.F., Tomé, C.N., Wenk, H.-R. (Eds.), Texture and Anisotropy: Preferred Orientation in Polycrystals and Their Effect on Material Properties. Cambridge University Press, Cambridge, pp. 283e324.

134. Trepmann, Stöckhert, B., Dorner, D., Moghadam, H.R., Küster, M., Röller, K., 2007. Simulating coseismic deformation of quartz in the middle crust and fabric evolution during postseismic stress relaxation e an experimental study. Tectonophysics 442, 83e104.

135. Treverrow, A., Budd, W.F., Jacka, T.H., Warner, R.C., 2012. The tertiary creep of polycrystalline ice: experimental evidence for stress-dependent levels of strainrate enhancement. Journal of Glaciology 58, 301e314.

136. Tullis, J., Christie, J.M., Griggs, D.T., 1973. Microstructures and preferred orientations of experimentally deformed quartzites. Geological Society of America Bulletin 84, 297e314.

137. Urai, J.L., Humphreys, F.J., Burrows, S.E., 1980. In-situ studies of the deformation and dynamic recrystallization of rhombohedral camphor. Journal of Material Science 15, 1231e1240.

138. Urai, J.L., Means, W.D., Lister, G.S., 1986. Dynamic recrystallization of minerals. In: Hobbs, B.E., Heard, H.C. (Eds.), Mineral and Rock Deformation; Laboratory Studies; the Paterson Volume, Geophysical Monograph, 36. American Geophysical Union, Washington, pp. 161e199.

139. Van der Veen, C.J., 1998. Fracture mechanics approach to penetration of surface crevasses on glaciers. Cold Regions Science and Technology 27, 31e47.

140. Van der Veen, C.J., Whillans, I.M., 1994. Development of fabric in ice. Cold Regions Science and Technology 22, 171e195.

141. von Mises, R., 1928. Mechanik der plastischen Formänderung von Kristallen. Zeitschrift für Angewandte Mathematik und Mechanik 8, 161e185.

142. Wakahama, G., 1964. On the Plastic Deformation of Ice. V. Plastic Deformation of Polycrystalline Ice. In: Low Temperature Science Series, A22, pp. 1e24 (In Japanese with English summary).

143. Wakahama, G., 1966. On the plastic deformation of single crystals of ice. In: Proceedings of the International Conference on Low Temperature Science, vol. 1. Institute of Low Temperature Science, Hokkaido University, Sapporo, Japan, pp. 292e311.

144. Wei, Y., Dempsey, J.P., 1994. The motion of non-basal dislocations in ice crystals. Philosophical Magazine A 69, 1e10.

145. Weikusat, I.S., Kipfstuhl, S., Faria, S.H., Azuma, N., Miyamoto, A., 2009. Subgrain boundaries and related microstructural features in EDML (Antarctica) deep ice core. Journal of Glaciology 55, 461e472.

146. Weikusat, I.S., Miyamoto, A., Faria, S.H., Kipfstuhl, S., Azuma, N., Hondoh, T., 2011. Subgrain boundaries in Antarctic ice quantified by X-ray Laue diffraction. Journal of Glaciology 57, 85e94.

147. Wenk, H.-R., 2000. Texture and Anisotropy: Preferred Orientation in Polycrystals and Their Effect on Material Properties. Cambridge University Press, Cambridge.

148. Weertman, J., 1983. Creep deformation of ice. Annual Review of Earth and Planetary Sciences 11, 215e240.

149. Wheeler, J., Mariani, E., Piazolo, S., Prior, D., Trimby, P., Drury, M., 2009. The weighted Burgers vector: a new quantity for constraining dislocation densities and types using electron backscatter diffraction on 2D sections through crystalline materials. Journal of Microscopy 233, 482e494.

150. Wilen, L.A., DiPrinzio, C.L., Alley, R.B., Azuma, N., 2003. Development principles, and applications of automated ice fabric analysers. Microscope Research Technology 62, 2e18.

151. Wilson, C.J.L., 1981. Experimental folding and fabric development in multilayered ice. Tectonophysics 78, 139e159.

152. Wilson, C.J.L., 1982. Texture and grain growth during the annealing of ice. Textures and Microstructures 5, 19e31.

153. Wilson, C.J.L., 1983. Foliation and strain development in ice mica models. Tectonophysics 92, 93e122.

154. Wilson, C.J.L., 1986. Deformation induced recrystallization of ice; the application of in situ experiments. In: Hobbs, B.E., Heard, H.C. (Eds.), Mineral and Rock Deformation; Laboratory

Studies; the Paterson Volume, Geophysical Monograph, 36. American Geophysical Union, Washington, pp. 213e232.

155. Wilson, C.J.L., Russell-Head, D.S., 1982. Steady state preferred orientation of ice deformed in plane strain at -1C. Journal of Glaciology 28, 145e160.

156. Wilson, C.J.L., Burg, J.P., Mitchel, J.C., 1986. The origin of kinks in polycrystalline ice. Tectonophysics 127, 27e48.

157. Wilson, C.J.L., Zhang, Y., 1994. Comparison between experiment and computer modelling of plane-strain simple-shear ice deformation. Journal of Glaciology 40, 46e55.

158. Wilson, C.J.L., Zhang, Y., 1996. Development of microstructure in the hightemperature deformation of ice. Annals Glaciology 23, 293e302.

159. Wilson, C.J.L., Russell-Head, D.S., Sim, H.M., 2003. The application of an automated fabric analyser system to the textural evolution of folded ice layers in shear zones. Annals Glaciology 37, 7e17.

160. Wilson, C.J.L., Russell-Head, D.S., Kunze, K., Viola, G., 2007. The analysis of quartz caxis fabrics using a modified optical microscope. Journal of Microscopy 227, 30e41.

161. Wilson, C.J.L., Peternell, M.A., 2011. Evaluating ice fabrics using fabric analyser techniques in Sørsdal Glacier, East Antarctica. Journal of Glaciology 57, 881e894.

162. Wilson, C.J.L., Peternell, M.A., 2012. Ice deformed in compression and simple shear: control of temperature and initial fabric. Journal of Glaciology 58, 11e22.

Citations

CHAPTER 1

Paul A Siratovich, Michael J Heap, Marlène C Villenueve, James W Cole, and Thierry Reuschlé, Physical Property Relationships of the Rotokawa Andesite, a Significant Geothermal Reservoir Rock in the Taupo Volcanic Zone, New Zealand, Doi:10.1186/s40517-014-0010-4.

CHAPTER 2

Upendra Singh Yadav and Vikas Mahto, Rheological Study of Partially Hydrolyzed Polyacrylamide-hexamine-pyrocatechol Gel System, doi: 10.1186/2228-5547-4-8.

CHAPTER 3

Z. Liu, L. Zhang, J. Jiang, C. Bian, Z. Zhang and Z. Gao, "Advancement of Hydro-Desulfurization Catalyst and Discussion of Its Application in Coal Tar," Advances in Chemical Engineering and Science, Vol. 3 No. 1, 2013, pp. 36-46. doi: 10.4236/aces.2013.31004.

CHAPTER 4

H. Lin, L. Ding, W. Deng, X. Wang, J. Long and Q. Lin, "Coating of Medical-Grade PVC Material with ZnO for Antibacterial Application," Advances in Chemical Engineering and Science, Vol. 3 No. 4, 2013, pp. 236-241. doi: 10.4236/aces.2013.34030.

CHAPTER 5

C. M. Julien, K. Zaghib, A. Mauger and H. Groult, "Enhanced Electrochemical Properties of LiFePO4 as Positive Electrode of Li-Ion Batteries for HEV Application," Advances in Chemical Engineering and Science, Vol. 2 No. 3, 2012, pp. 321-329. doi: 10.4236/aces.2012.23037.

CHAPTER 6

Erler, J. , Leistner, T. and Peuker, U. (2014) Application of a Particle Extraction Process at the Interface of Two Liquids in a Drop Column—Consideration of the Process Behavior and Kinetic Approach. Advances in Chemical Engineering and Science, 4, 149-160. doi: 10.4236/aces.2014.42018.

CHAPTER 7

Shike Zhang, Shunde Yin, Determination of in situ stresses and elastic parameters from hydraulic fracturing tests by geomechanics modeling and soft computing, Journal of Petroleum Science and Engineering, Volume 124, December 2014, Pages 484-492, ISSN 0920-4105, http://dx.doi.org/10.1016/j.petrol.2014.09.002.

CHAPTER 8

Christopher J.L. Wilson, Mark Peternell, Sandra Piazolo, Vladimir Luzin, Microstructure and fabric development in ice: Lessons learned from in situ experiments and implications for understanding rock evolution, Journal of Structural Geology, Volume 61, April 2014, Pages 50-77, ISSN 0191-8141, http://dx.doi.org/10.1016/j. jsg.2013.05.006.

Index